SpringerBriefs in Earth Sciences

For further volumes:
http://www.springer.com/series/8897

João Pais · Pedro P. Cunha
Diamantino Pereira · Paulo Legoinha
Rúben Dias · Delminda Moura
António Brum da Silveira
J. C. Kullberg · J. A. González-Delgado

The Paleogene and Neogene of Western Iberia (Portugal)

A Cenozoic Record in the European Atlantic Domain

 Springer

João Pais
CICEGe, Dep. Ciências da Terra
Faculdade de Ciências e Tecnologia
Universidade Nova de Lisboa
2829-516 Caparica
Portugal
e-mail: jjp@fct.unl.pt

Pedro P. Cunha
Earth Sciences Department
IMAR-Marine and Environmental Centre
University of Coimbra
3000-272 Coimbra
Portugal
e-mail: pcunha@dct.uc.pt

Diamantino Pereira
Centro de Geologia da Universidade do
    Porto
Centro de Ciências da Terra da Universidade
    do Minho,
Campus de Gualtar
4710-057 Braga
Portugal
e-mail: insuad@dct.uminho.pt

Paulo Legoinha
CICEGe, Dep. Ciências da Terra
Faculdade de Ciências e Tecnologia
Universidade Nova de Lisboa
2829-516 Caparica
Portugal
e-mail: pal@fct.unl.pt

Rúben Dias
Laboratório Nacional de Energia e Geologia
    (LNEG/LGM)
Dep. Geologia, Estrada do Zambujal, Apart. 7586
2721-288 Alfragide
Portugal
e-mail: ruben.dias@lneg.pt

Delminda Moura
Marine Research Centre
Universidade do Algarve (CIMA-UALG)
Campus de Gambelas
8000-139 Faro
Portugal
e-mail: dmoura@ualg.pt

António Brum da Silveira
Laboratório de Tectonofísica e Tecnónica
    Experimental/Instituto Dom Luís (Lattex/IDL)
Faculdade de Ciências
Universidade de Lisboa
1749-016 Lisboa
Portugal
e-mail: antonio.brum@fc.ul.pt

J. C. Kullberg
CICEGe, Dep. Ciências da Terra
Faculdade de Ciências e Tecnologia
Universidade Nova de Lisboa
2829-516 Caparica
Portugal
e-mail: jck@fct.unl.pt

J. A. González-Delgado
Departamento de Geología
Facultad de Ciencias
Universidad de Salamanca
37008 Salamanca
Spain
e-mail: angel@usal.es

ISSN 2191-5369
ISBN 978-3-642-22400-3
DOI 10.1007/978-3-642-22401-0
Springer Heidelberg Dordrecht London New York

e-ISSN 2191-5377
e-ISBN 978-3-642-22401-0

Library of Congress Control Number: 2011936136

Cover design: eStudio Calamar, Berlin/Figueres

Printed on acid-free paper

Springer is part of Springer Science+Business Media (www.springer.com)

Contents

The Paleogene and Neogene of Western Iberia (Portugal): A Cenozoic Record in the European Atlantic Domain

1 Introduction

The Portuguese mainland, located in western Iberia, represents a key area for understanding the evolution of the European Atlantic margin during the Cenozoic and the establishment of relations with the Mediterranean, in particular through the transition area between those two domains which is well documented in the Algarve region.

Iberia, positioned between the Eurasian and African plates, has been moving eastward since the Triassic due to the progressive opening of the Atlantic Ocean. After the Late Cretaceous, the Triassic extensional tectonic regime was replaced by compressive phases due to collision between the two plates, leading to the development of sedimentary basins, generally oriented E–W to NE–SW. During the Cenozoic, Iberia underwent intense intraplate compressive deformation that caused lithospheric folding (Cloetingh et al. 2002; Tejero et al. 2010). Maximum compression was generally oriented along a N–S orientation (Vegas 2006); however, since the Late Miocene, Iberia has rotated to the NW–SE (Ribeiro et al. 1996; De Vicente et al. 2008, 2011; Kullberg et al. 2011b).

The Portuguese Cenozoic basins—the Mondego, Lower Tejo, Guadiana, Alvalade and Moura basins—began to be filled in the early Middle Eocene (Figs. 1 and 2). These basins correspond to NE–SW elongated depressions separated by passive doorways. Until the middle Tortonian, their evolution was marked by a slow and gradual erosion of the Hesperian Massif, under continuous tectonic deformation and climatic conditions (semi-arid to subtropical climate with long dry season) that favoured the planation of the basement and the transport of feldspathic sands to the basins.

In Portugal mainland, the peak of the Alpine compression occurred in the mid-Tortonian (about 9–9.5 Ma; Cunha 1992; Cunha et al. 2000), when major mountain ranges, such as the Central Portuguese Cordillera (2000 m altitude) and the Western Mountains, began to be uplifted. In the Late Miocene and Zanclean,

J. Pais et al., *The Paleogene and Neogene of Western Iberia (Portugal)*, SpringerBriefs in Earth Sciences, DOI: 10.1007/978-3-642-22401-0_1, © João Pais 2012

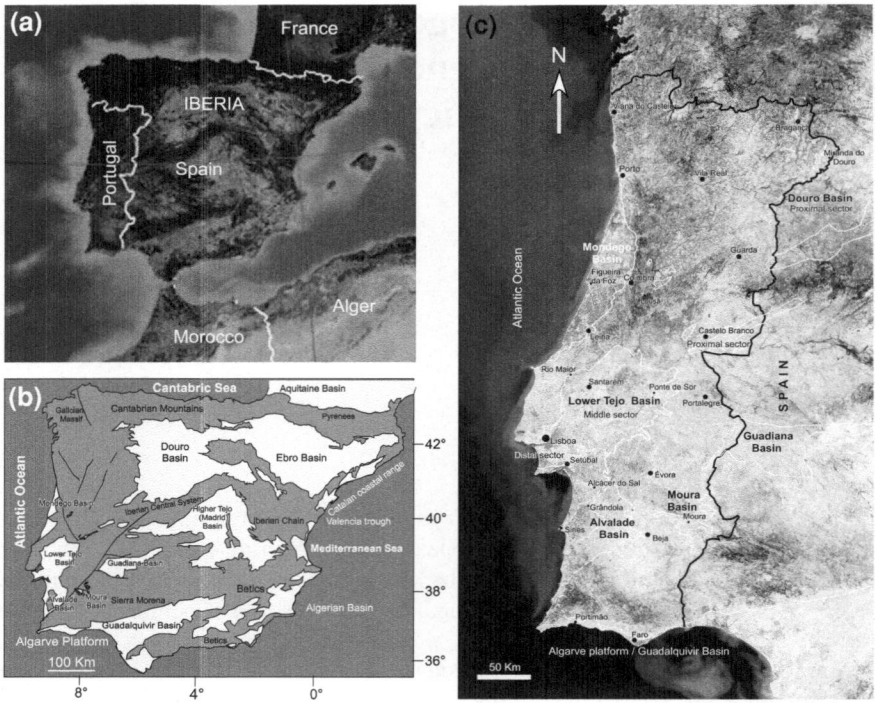

Fig. 1 **a** General location of the Portuguese mainland in the Iberia Peninsula; **b** Schematic map of the main Iberia morphostructural units and Cenozoic basins; **c** Locations of Cenozoic deposits on the Portuguese mainland

under a hot climate with strong inter-seasonal differences, sedimentation was expressed by endorheic and alluvial fans at the foots of scarps of active, mainly reverse, faults oriented NE–SW to ENE–WSW, NNE–SSW and WNW–ESE. In the Pliocene, the hot climate of the Piacenzian became very wet. An exorrheic hydrographic network developed, predecessing the present one. Broad river valleys were configured in mountainous areas and numerous catchments were developed in inland endorheic basins. In the Gelasian, the climate became colder; the continuation of regional tectonic uplift and periods of low sea level were determinants of the progressive evolution of river systems (strong incision, regressive erosion and stream piracy).

Studies of the offshore areas of the West Iberian Margin (WIM) (Mougenot 1989; Pinheiro et al. 1996; Alves et al. 2003, 2011) and southern Portugal (Lopes 2002; Lopes et al. 2006, 2008a, b; Lopes and Cunha 2007; Terrinha et al. 2011) have allowed the infilling geometries of these areas to be understood and their correlatives identified in the onshore sedimentary unconformities defined by Cunha (1992).

The unconformities recorded in the Iberian Cenozoic sedimentary basins can be correlated with episodes of compression between Iberia, Eurasia and Africa. The sedimentary record of the major Cenozoic Portuguese basins includes continental units in the interior which, progressively, pass to marine sedimentation in areas closer to the present coastline, reflecting various paleogeographic, tectonic, climatic and eustatic events.

Allostratigraphic units have been defined in the WIM, and the high-resolution sequential evolution of the Neogene distal sector of the Lower Tejo Basin has been characterized (Cunha 1992; Cunha and Reis 1992; Reis et al. 1992a, b; Antunes et al. 1999b, 2000; Legoinha 2001, 2008; Cunha and Martins 2004; Pais 2004; Cunha et al. 2008b, 2009).

In the Trás-os-Montes and Nave de Haver regions, small depressions preserve continental deposits, usually not very thick and with little chronostratigraphical control. It is considered that the sedimentary record matches the proximal sector of the Cenozoic Douro Basin's western margin (Fig. 1). The basal filling of the Vilariça-Longroiva strike-slip basins (Cunha and Pereira 2000) has been correlated with the Nave de Haver arkoses (Paleogene to Miocene?). The Neogene sediments that fill these basins are similar to the sediments from the easternmost sector of Trás-os-Montes, filling excavated paleovalleys in the Variscan substrate and with paleodrainage oriented with respect to the Cenozoic Douro Basin (Pereira 1997, 2006).

In the Mondego, Lower Tejo, Alvalade and Moura basins, sedimentation continued throughout the Cenozoic. The Lower Tejo Basin has been affected by several marine incursions in the Miocene (until the middle Tortonian) and in the Pliocene; the Alvalade basin in the terminal Miocene (Messinian) and the Mondego basin were only ephemerally occupied in the early Late Pliocene.

The Moura Basin comprises several tectonic sub-basins with a thin sedimentary record of continental origin, mainly composed of Miocene to Pleistocene alluvial fan deposits, related to tectonic activity, and preserved fluvial deposits associated with a Miocene E to W oriented paleodrainage.

In the Algarve region, a marine platform developed during the Cenozoic on which deposits from the Lower and Middle Miocene were subsequently accumulated. The units of the upper Tortonian and lower Messinian mainly correspond to the western infilling sector of the Guadalquivir Basin.

2 Tectonic Evolution of Western Iberia During the Cenozoic

2.1 The Regional Context

The Iberian Plate (the Iberian microcontinent or simply Iberia) underwent a complex tectonic evolution during the Alpine Orogeny. Rifting during the Mesozoic was the dominant process: in the western margin it was related to the first phases of the opening of the North Atlantic, and in the eastern and southern margins to the westward propagation of the Thetys Ocean. This tectonic context led to the formation of several extensional and transtensional basins in all margins of Iberia.

After the beginning of the breakup of Pangea, Iberia became bounded by major lithospheric structures, namely the Charlie Gibbs Fracture Zone to the north, the Açores-Gibraltar Fracture Zone to the south (Verhoef and Srivastava 1989), the Atlantic Rift to the west and the Tethyan Rift to the south and southeast. Because of this pattern, Iberia either acted as an independent microplate or moved attached alternately to the Eurasian or African plates during the Alpine cycle (Klitgord and Schouten 1986; Srivastava et al. 1990; Roest and Srivastava 1991). The former implies continuous changes in the position of the plate boundaries among the three plates; the latter implies shifts in the plate boundaries between Eurasia and Africa, Iberia being part of one or the other. Aspects of the evolving tectonic configuration remain contentious, including how frequently, and when, such changes in position occurred (e.g., Malod 1989; Vega 1989; Srivastava et al. 1990; De Vicente et al. 2005).

Based on aeromagnetic measurements, analysis of seafloor magnetic anomalies, regional geological data and analogue models, authors have used both qualitative and quantitative approaches to model the tectonic history of Iberia. Some of these studies have been more focused on interplate kinematics (e.g., Schettino and Scotese 2002) and others on intraplate deformation (e.g., De Vicente et al. 2005; 2011). Nevertheless, there is a general consensus among the various investigators that Iberia has been subject to compression since the end of the Mesozoic, due to a N–S convergence between Africa and Eurasia (Dewey et al. 1989; Rosenbaum et al. 2002). At a broader scale, in the western Tethys, convergence between Africa and Eurasia, which started in the Early Cretaceous, triggered further deformation within the dispersed continental fragments and the formation of backarc basins in the active margins (Schettino and Turco 2011); this led to an increase in the number of tectonic elements that were moving independently in the western Tethyan region during the Late Cretaceous and the Cenozoic.

The consequent shortening and related morphogenesis of Iberia is therefore mainly related to the N–S oriented Oligocene to Late Miocene convergence between Africa and Europe (the Betic Orogeny), and to the approximately N–S oriented shmax transmitted stresses (Andeweg et al. 1999; Escuder-Viruete et al. 2001; De Vicente et al. 2009). After Cunha and Reis (1995) these N–S SHmax transmitted stresses had already reached significant levels in the middle Campanian. During the Late Cretaceous to Early Tertiary, Iberia moved northward against the Eurasian Plate, forming the Pyrenean orogen. This caused the some earlier extensional structures to become inverted and shortened, forming two fold-thrust belts and associated foreland basins (Muñoz 1985; Coney et al. 1996). The crustal shortening and resulting intense intraplate deformation were responsible for a drastic change in the continental topography of Iberia (De Vicente et al. 2004, 2011). From an average elevation close to sea level until the end of the Cretaceous (Cunha and Reis 1995; Dinis et al. 2008), the planation surface became uplifted to between 100 and 600 m, with E–W oriented mountain belts rising to approximately 2000 m (De Vicente et al. 2011).

Several authors (e.g., Cloetingh et al. 2002; Fernández-Lozano et al. 2011; De Vicente et al. 2011) suggest that lithospheric folding was the primary response to large-scale shortening of Iberia during the Betic Orogeny, which was probably enhanced by the reactivation of inherited Variscan faults. Folding is associated with the formation of predominantly pop-up type mountain ranges in the brittle crust and thickening of the ductile layers in the synforms of the buckle folds by flow (Fernández-Lozano et al. 2011). Therefore, the mountain ranges are represented by upper crustal pop-ups forming the main topographic relief. In consequence, sedimentary basins were formed in areas that were subsiding relative to the ranges, in which sediments accumulated to form thick successions reaching up to 3500 m in the case of the Madrid Basin (Lanaja 1987).

The large number of basins of various sizes that developed during the Cenozoic are bounded by topographic highs marked by thrusts or strike-slip faults (De Vicente et al. 2011). The main uplifted zones in Iberia are:

(1) In the north, the Cantabrian-Pyrenean mountains that extend to the west to the Galicia Massif and the northwestern Portuguese mountains;
(2) In the centre of Iberia, the Iberian Chain and the Central System, extending to the west to the Maciço Calcário Estremenho and the Estremadura Spur offshore of the West Iberian Margin; and
(3) In the south, the Sierra Morena-Betic cordillera.

The ranges located in these three structural uplifts were formed by crustal buckling and contain the main thrusts that are bounded by left- and right-lateral strike-slip deformation belts (De Vicente et al. 2011). The main thrusts are located in the brittle upper crust and are the superficial expression of deep litospheric folding. For example, the ENE–WSW oriented Central System is considered to be a deep lithospheric anticline bordered by two synclines; these features are responsible for the development of the Douro Basin, to the north, and the Tejo Basin to the south (Cloetingh et al. 2002; De Vicente et al. 2004; Vegas 2005; Martín Velázquez and Elorza 2007), which are the most important Cenozoic Basins located in the northern and southern sectors of the microplate. Therefore, the Central System would correspond to a thick-skinned pop-up structure as proposed by Ribeiro et al. (1990), but related to the displacement to the north of a deep basement detachment from the Betic Cordillera to upper crustal structures, such as the Central System.

The Iberian microplate is crosscut by a complex set of major fractures, inherited mainly from the Variscan Orogeny. They were later reactivated in an extensional context during the Mesozoic and controlled the formation and evolution of basins in the Iberian margins. During the Cenozoic compression, they were also reactivated: the E–W to NE–SW fractures as thrusts, the NNE–SSW as left-lateral strike-slip faults, and the NW–SE as right-lateral strike-slip faults. Considering that the entire Iberian lithosphere was being regularly deformed by an approximately N–S Shmax orientation during the Cenozoic, basins conditioned by that set of faults developed in a variety of local tectonic settings, from thrust basins to strike-slip transtensional and transpressional basins. De Vicente et al. (2011) demonstrated that very different deformation styles and associated basin infilling features have been conditioned by a differentiated tectonic response to phases of relatively homogeneous tectonic stresses on an intraplate scale.

Other classifications of the Iberian Cenozoic basins have been undertaken:

(1) Friend and Dabrio (1996) and Gibbons and Moreno (2002) among other authors consider either their relationship with the Pyrenean or the Betic orogenies, or their intraplate location;
(2) Civis (2004) have classified the Iberian basins according to the type of basement-Variscan or Mesozoic.

De Vicente et al. (2011) combined a tectonic analysis with a calculated ratio between the length of the strike-slip faulting (ss) and the length of the dip-slip (thrust) (ds) faulting affecting the basin borders [r = (ss)/(ds)]. Their conclusions included:

(1) Transtensional basins (pull-apart or releasing bends) have high r values (r = 3–5), meaning that strike-slip faults are the mainly bordering faults
(2) In transpressional basins, both strike-slip and dip-slip bordering faults condition the basins' development (1 > r > 0.8 and r = 3, respectively), and the open ramp basins show slightly lower values of r (=0.6) and correspond to the main Iberian Cenozoic basins; and
(3) the open ramp basins showed slightly lower values close of r (0.6) and they correspond to the main Iberian Cenozoic basins.

2.2 Cenozoic Inversion in Western Iberia

2.2.1 The Late Variscan to Mesozoic Heritage

West Iberia (Fig. 1) is a typical example of a non-volcanic rift margin that experienced continental extension, traditionally considered to have occurred from the Upper Triassic (e.g., Wilson 1989; Mauffret et al. 1989) to the latest Early Cretaceous (e.g., Pinheiro et al. 1996; Rasmussen et al. 1998; Kullberg 2000). During this stage of continental stretching, which led to the oceanization of the North Atlantic, the process of continental breakup progressively migrated from the south to the north, along the margin (Driscoll et al. 1995; Kullberg 2000, 2006b; Alves et al. 2006). In the west Iberia margin, several extensional basins evolved during the Mesozoic; they are, from south to north, the Alentejo Basin, the Lusitanian Basin; the Peniche Basin, the Porto Basin and the internal and external Galicia basins.

The Algarve Basin localized in the southwest Iberia in a transtensional shear zone that was part of the Neo-Tethys rift system formed as a consequence of the eastward drift of Africa with respect to Eurasia (Dewey et al. 1989; Malod and Mauffret 1990; Srivastava et al. 1990). However, the transcurrent deformation associated with the differential movement of the two plates must have been accommodated in domain external to the Algarve Basin, between the Guadalquivir Bank and northern Africa, because most of the movement on the NE–SW to ENE–WSW oriented extensional faults of the Algarve Basin is dip-slip (Terrinha 1998; Terrinha et al. 2002).

During the Mesozoic extensional period, western basins were mainly controlled by deep normal faults oriented NNE–SSW to N–S and NW–SE, corresponding to the reactivations of inherited Variscan faults. These basins are approximately continuous in space and time, and although the sedimentary records of the basins differ, a thick evaporitic sequence of Hettangian age underlies almost all of them. The basins are separated by major transfer faults oriented from NE–SW to ENE–WSW (from south to north, the Arrábida Fault, the Nazaré Fault and the Aveiro Fault) that developed important roles during the Alpine compression, mainly in the Cenozoic.

Another major inherited Variscan discontinuity is the Messejana Fault, which separates the West Iberian Margin basins from the Algarve Basin in the southwest corner of Iberia. It is intruded by a dike with the composition of a typical continental tholeiitic basalt, which crops out for some 530 km with a thickness varying from 5 to 300 m. The dike was emplaced at around 200 Ma in the northernmost sector of the Central Atlantic Magmatic Province (Cebriá et al. 2003; Martins et al. 2008) to where the Algarve Basin pertained in the Lower Jurassic. It is therefore a major crustal fault that has had a complex tectonic history (Schermerhorn et al. 1978; Schott et al. 1981) as an initial Late Variscan sinistral strike-slip fault, later reactivated as a transtensional fault allowing the emplacement of Triassic-Jurassic magmas, and again reactivated as a transtensional sinistral strike-slip fault during the Cenozoic (Cebriá et al. 2003).

No magmatism is registered in the geological record to the north of the Messejana Fault for the Triassic-Jurassic extension in the West Iberian Margin. However, the onshore sector of the West Iberian Margin was the locus of other cycles of magmatic activity during the Mesozoic, the most voluminous of which was of alkaline composition and occurred between 70 and 100 Ma. Miranda et al. (2009) argue that this cycle took place in a post-rift environment, during a 26°–35° counter-clockwise rotation of Iberia (e.g., Moreau et al. 1997; Juarez et al. 1998; Márton et al. 2004; Martins et al. in Kullberg et al. 2011a, b) and initiation of the alpine compression. This allowed the emplacement of the Tore (offshore), Sintra, Sines, Monchique and Guadalquivir (offshore) subvolcanic complexes along NNW–SSE to NW–SE oriented dextral, deep-seated faults to a total length of around 600 km.

The Mesozoic extensional tectonic regime lasted for about 100 million years, and produced distinct crustal domains of thinned continental crust, from the oceanic domain in the west to the continental hinterland in the east (Afilhado et al. 2008; Alves et al. 2009). Afilhado et al. (2008) identified four main crustal domains from their analysis of coincident, multi-channel, near-vertical and wide-angle reflection data sets:

(i)   Continental (east of 9.4°W);
(ii)   Continental thinning (9.4°–9.7°W); and
(iii) Transitional (9.7°–10.5°W); and
(iv)   Oceanic (west of 10.5°W).

In the continental domain, the complete crustal section of slightly thinned continental crust is present, demonstrating that the thinning was not completely recovered by the Cenozoic compression (Afilhado et al. 2008).

Most of the Mesozoic basins are localized in the first and second domains, meaning that most of them are located offshore in the present day. With the exception of about a third of the Lusitanian Basin that crops out in the central part of the West Iberia Margin, and also part of the Algarve Basin in the soutwest, all the other basins are located in the platform, preserving a thick Meso-Cenozoic stratigraphic succession that has been thoroughly imaged by seismic methods and calibrated by deep drilling.

2.2.2 The Cenozoic Compression

Cenozoic Compression

The Alpine foreland in Portugal was deformed by compressional tectonics principally during the Miocene. This orogeny is registered in the west and southwest thinned Mesozoic margins of Iberia filled by thick sedimentary packages, and also in the Variscan igneous and metamorphic thick basement in the neighbouring hinterland. This is a key factor that conditions the style of inversion of the previous basins and their controlling structures, and also the loci of newly formed basins.

In the Lusitanian and Algarve basins, the Meso-Cenozoic cover was deformed, mainly during the Miocene (Ribeiro et al. 1990). The same authors consider that the pre-Mesozoic basement is also involved in Alpine thick-skinned structures in western Iberia. Late Variscan basement-cutting strike-slip faults striking between ENE–WSW and NE–SW were reactivated as reverse faults and both NW- and SE-verging thrusts. The geometry of lateral ramps and duplexes shows the direction of transport of the basement rocks onto the Miocene cover. The largest structure thus formed was the Central System, which was thrust over Miocene sediments both to the northwest and to the southeast (Cabral 1995; Ribeiro et al. 1990). The SE-verging, NW-dipping Ponsul fault, which bounds the Central System on its southeast end side, flattens at depth and converges with the NW-verging, SE-dipping Seia-Lousã fault (the eastern branch of the Nazaré fault in the basement), which bounds the Central System to the northwest, thus identifying this structure as a basement pop-up (Ribeiro et al. 1990).

The main Shmax transmitted stresses estimated for the Portuguese area of Iberia do not substantially differ from the N–S direction reported for the remainder of

Iberia; based on inversion structures in the Lusitanian Basin, Ribeiro et al. (1996) and Kullberg et al. (2000) determined a NNW–SSE direction for Shmax related to the Oligocene–Miocene compressive episode. The main faults reactivated during the alpine compression were (Fig. 2):

(1) The NE–SW to ENE–WSW oriented structural pattern as thrusts;
(2) The NNE–SSW as sinistral strike-slip faults; and
(3) The NW–SE as dextral strike-slip faults.

In the onshore area, except in the Mondego basin that contains a relevant Upper Campanian-Ypresian sedimentary succession, the early compressive epi-sode considered by several authors as Late Cretaceous-Early Cenozoic is not well documented since, the sedimentary record comprises mainly continental environments, and it is thin and discontinuous and finally suffered intense ero-sion resulting from uplift of the margin. As in the offshore area, the sedimentary package is fairly complete, although sometimes not controlled by dating from boreholes. Similarly as the compressive events identified by Cunha (1992), Pereira et al. (2010), based on the interpretation of extensive 2D seismic surveys performed in the Alentejo Basin (offshore) and on the identification of chief erosional surfaces, consider three distinct compressive events. The first was initiated in the Late Cretaceous, likely associated with the uplift of the hinterland by intrusion of large igneous bodies and post-rift thermal differential subsidence across the margin. The second was interpreted as recording a mid-Eocene compressive phase, expressed in the distal margin by incipient folding. The third was of Oligocene to mid-Miocene age, expressed in the proximal margin by localized folding and reverse faulting, which were rooted from a deep detach-ment in the viscous evaporitic sequence of the Hettangian. While in the proximal margin the inversion is marked by deep incision of plaeocanyons interpreted as resulting from uplift (Mougenot et al. 1989; Mougenot et al. 1989), in the distal margin shortening is expressed by broad anticlines, reverse faults and "piggy-back" thrusting. Both folding and faulting in this sector are broadly aligned WSW–ENE, suggesting that the chief compressional forces were likely domi-nated by the Miocene direction of compression (see above). The first two compressional episodes identified by Pereira et al. (2010) coincide with the Late Cretaceous-Paleogene collision of the Pyrenean Orogeny, and the third is related to the Betic Orogeny.

Eocene deformation is also represented in the Galicia basins (Pinheiro et al. 1996) but the Miocene deformation dominates the rest of the margin, where it may have overprinted the former compressional episode. The Pyrenean Paleogene compression is progressively less evident from the north to the south, i.e., from the Porto to the Lusitanian and Peniche basins to the Alentejo Basin (Alves et al. 2003), and the trend of attenuation of deformation from the inner orogen located in the NNE and of Iberia (the Cantabrian-Pyrenean structural uplift, after De Vicente et al. 2011) to the external zones is clear in this zonation.

The Algarve Basin has a complex Cenozoic tectonic evolution with respect to its geometry and particular processes involved. The shallow water sediments of the

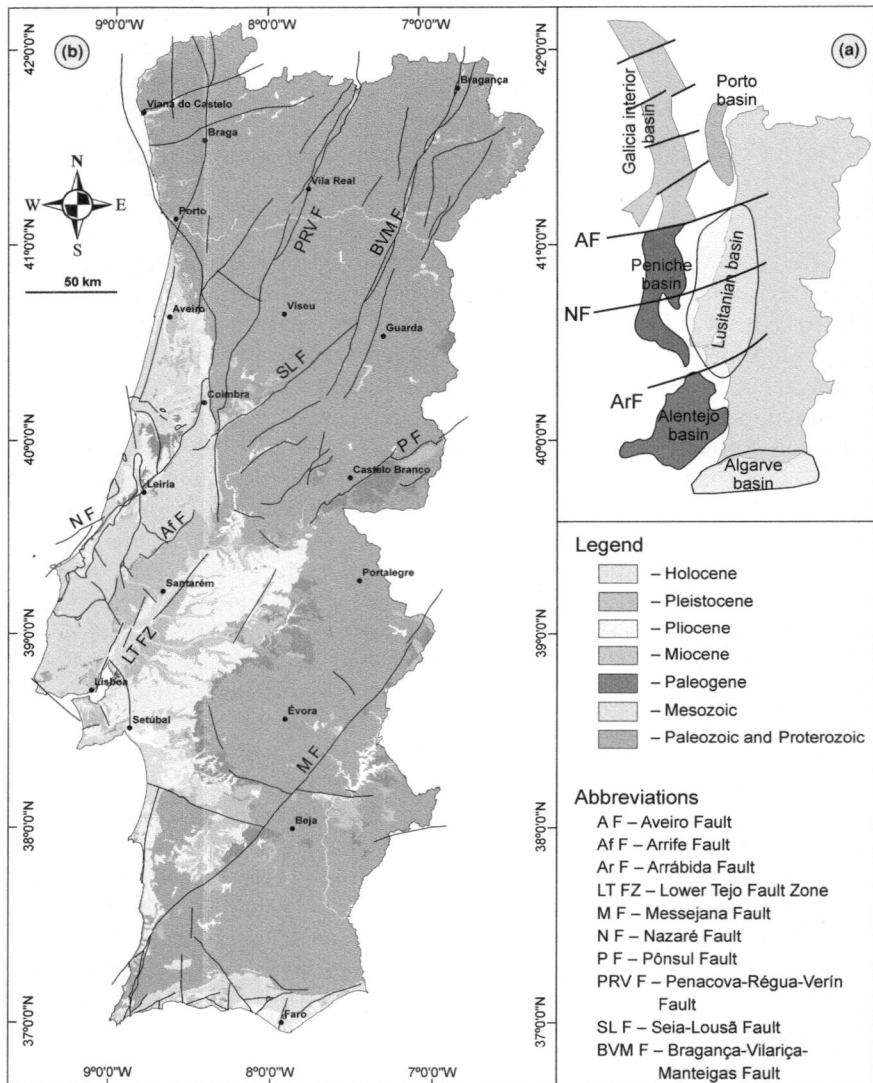

Fig. 2 a General map of Portugal with the main mesozoic basins. **b** Generalised map of the Cenozoic basins and units of Portugal (based on LNEG 2010; Cabral and Ribeiro 1988; Cabral 1995)

Paleogene, which are preserved only in the offshore area (Lopes et al. 2006), are related to the same compressional episode that shortened and inverted the Mesozoic basin. Terrinha (1998) and Terrinha et al. (1998, 2006) consider that the Neogene exhibits negligible deformation and is difficult to characterize in terms of geometry and kinematics. However, based on seismic profile interpretation calibrated by borehole stratigraphy, Lopes et al. (2006) present a different interpretation involving a complex Cenozoic inversion history. Those authors propose a moderate to intense phase of shortening during the Paleogene that provoked strong tectonic inversion, with uplift, folding, thrusting and the generation of an important unconformity. The E–W-trending Mesozoic half-graben became progressively inverted, in thick-skinned style tectonics, and during a period of apparent quiescence between the Lutetian and the Oligocene the reactivation of basement-related structures generated widespread halokinesis (Lopes et al. 2006). After this, from the Aquitanian to middle Tortonian, regional compression appears to have increased. A salt/fault-controlled thin/thick-skinned subsidence influenced both the thickness and lateral distribution of sediments.

It is in the onshore sector of central Portugal, mainly in the Mesozoic cover of the Lusitanian Basin, that the Cenozoic tectonic inversion is better expressed. Due to an important and long lasting uplift of the entire region, from the Late Cretaceous to the Paleogene, the sedimentary record and thus tectonic structures are almost non-existent. The exact causes, thus the timing and relevance of the several phases of this uplift, are not yet well known, but can be related to one or more of the following processes:

(1) Late Cretaceous magmatism that could have been triggered by crustal thickening by underplating due to astenospheric upwelling (Matton and Jébrak 2009);
(2) The Pyrenean Orogeny; and
(3) The counter-clockwise movement of Iberia, eventually related to the onset of the Pyrenean Orogeny.

Some small basins and local structures are, however, preserved and show deformation compatible with the Pyrenean Orogeny or locally induced deformation by halokinesis, as occurred in widely in the marginal Meso-Cenozoic basins. Regarding the latter, Terrinha et al. (1996) refer to small basins formed during the first tectonic inversion, such as at Leiria, Alcobaça and Rio Maior, which are associated with sagging due to salt withdrawal and salt extrusion and welding. Other small structures such as the Alcanena syncline are good examples of deformation compatible with the Pyrenean Orogeny; they exhibit folded Late Paleogene units unconformably overlying a tighter syncline affecting both Lower and Upper Cretaceous units.

During the Mesozoic extension, a thick-skinned tectonic style prevailed, which was associated with the normal faults of basin borders and inner- and half-graben structures; in some locations, at very high structural levels, a thin-skinned style can be found (Kullberg 2000; Kullberg et al. 2006a). However, during compression related to the Betic orogeny, the Meso-Cenozoic cover was deformed in a thin-skinned style with décollement in the Hettangian evaporites (Ribeiro et al. 1990a; Kullberg 2000; Alves et al. 2002; Kullberg et al. 2006b).

The most important reliefs located in the central western part of Iberia are of Lower Miocene age, demonstrating a progressive propagation of the deformation from the front of the Oligocene–Miocene Betic chain to the foreland basins, in our case localized approximately 400 km to NW of the Betic Orogen. One of the more important is Arrábida, a small chain elongated ENE–WSW that represents the most elegant example of thin-skinned Alpine tectonics in western Iberia (Ribeiro et al. 1990b). In this chain, it has proved possible to accurately date the paroxism of the Early Miocene compressional event: using isotopic methods, Antunes et al. (1995) dated the episode as being within the Burdigalian (17.5–16.5 Ma).

The most important topographic anomaly over the Portuguese continent is an alignment of reliefs from about 500–2000 m, which extends for more than 250 km and continues into the Iberian Meseta. It corresponds to the Portuguese branch of the inverted Central System (sensu De Vicente et al. 2011) formed by the Estrela-Seia-Lousã (in the basement) and the Aire-Candeeiros-Montejunto (in the Mesozoic cover) mountain systems also with a very impressive topographic signature in Estremadura Spur located in the platform. The inverted structures in the Mesozoic sedimentary cover (the Lusitanian Basin) are anticlines associated with thrust faults, pertaining to the complex system formed by the Arrife and Lower Tejo faults that trend from NE–SW to E–W, including the offshore Estremadura Spur (Roque et al. 2009).

The inversion tectonics also triggered the formation of indentors, which are associated with sub-basin geometries, either by the reactivation of extensional structures or by the generation of basins during compression. Sectors along the borders of diapirs (e.g., Caldas da Rainha) or in the eastern sector of the Arrábida Chain are examples of these thin-skinned structures. They result from constriction related to tectonic movement directed NW–SE in areas where NNE–SSW oriented faults intersect with those trending E–W to NE–SW. In the Caldas da Rainha example, the indentor is very shallow, and is detached at the contact between the upper evaporitic unit and carbonates in the top. In the Arrábida Chain case, the indentor is formed from a basement buttress in the footwall of the intersecting faults (Kullberg et al. 2000, 2006b); the faults correspond to Mesozoic bordering normal faults of the Lusitanian Basin that have not recovered the offset due to extension during the Cenozoic inversion (this is another demonstration of the prevailing thin-skinned style). Other resulting structures are the imbrication of thrusts in a "piggy-back" geometry in the south Arrábida verging chain. The propagation of thrusting is from south to north, and represents the final pulse of Tortonian age.

Several authors consider that the Arrábida event is the most important compressive episode responsible for the main elements of relief both in the Portuguese offshore area and also in the remainder of Iberia. De Vicente et al. (2011) concluded that towards $\sim$9.5 Ma (late Vallesian, middle Tortonian) a very important paleogeographic change occurred in all the Iberian basins (apart from the extensional ones). This drastic paleogeographic change during the middle Tortonian reflects a large-scale tectonic event induced by the climax of compression in the western part of the Central System.

Like the other Iberian basins the Portuguese Cenozoic basins were classified by De Vicente et al. (2011) after the tectonic context under which they were formed and evolved. The smaller basins are related to transpressional and transtensional tectonic regimes. Basins integrated in the first system are the following: Chaves, Vila Real, Besteiros and Mortágua associated with the Régua-Verin left-lateral strike-slip fault and Bragança (refered as Braga), Vilariça and Longroiva associated with the Vilariça left-lateral strike-slip fault. Basins developed under transtensional conditions (pull-apart basins) are the following: Lucefeci, Montoito, Odemira (refered as Dermira) and Aljezur associated with the Messejana Fault that was reactivated during the Cenozoic compression as a hight angle (relative to the Shmax) left-lateral strike-slip fault. The other basins evolved conditioned by thrusts, mainly in the Central System and represent the larger in the Portuguese margin: the Mondego and the Lower Tejo basins, but some tectonic depressions as the Sarzedas and Ponsul sub-basins. Cenozoic sub-basins also developed in the inverted areas of the Lusitanian, the Algarve and offshore basins, mostly conditioned by local halokinesis triggered in different times by different processes, including Cenozoic compression or by thin-skinned structures.

The organization of these dispersed basins is here simplified since most of the basins cited by De Vicente et al. (2011) in the western part of Iberia are, in fact, sub-basins of larger basins that were exhumed presumably from the Late Neogene to the Present. After Casas and De Vicente (2009), isostatic adjustment resulting from crustal thickening, erosion during the endorheic–exorheic drainage transition and the lithospheric folding process in southern Iberia (also resulting in elevation of intra-basinal areas) were different results of the decoupling induced by the emplacement of the Alboran domain in southern Iberia during the Europe–Iberia–Africa convergence. So we prefer to consider a broader genetic approach (Fig. 1) incorporating the small basins mainly those aligned in main faults as the Régua-Verín and Vilariça in the North and the Messejana in the South in the Douro and the Moura (sl) basins, respectively. The small basins (sub-basins) correspond, in our point of view, to tectonically preserved outcrops in the vicinity of transtensional/transpressional that crosscut those large domains. For the Moura basin (eventually extend during the Miocene further SW) field data and mapping (e.g., LNEG 2010) do not confirm in the portuguese sector of the Messejana fault the existence of the roughly N–S orientated (probably normal faults) refered by De Vicente et al. (2011). The main sedimentary organization of the basin is clearly conditioned by E–W thrusting.

During the Pliocene there was some tectonic stability. In the Late Pliocene and Quaternary, the tectonic regime became compressive with a WNW-ESE to NW-SE direction of maximum compression (Ribeiro et al. 1996). This tectonic instability was probably due to changing conditions at the Azores-Gibraltar tectonic plate boundary (Zitellini et al. 2009; Terrinha et al. 2009) and at the southwest Portuguese continental margin; these changes probably represented a transition from a passive to a compressive active margin, possibly associated with the nucleation of a subduction zone (Cabral and Ribeiro 1989; Ribeiro et al. 1996; Ribeiro 2002).

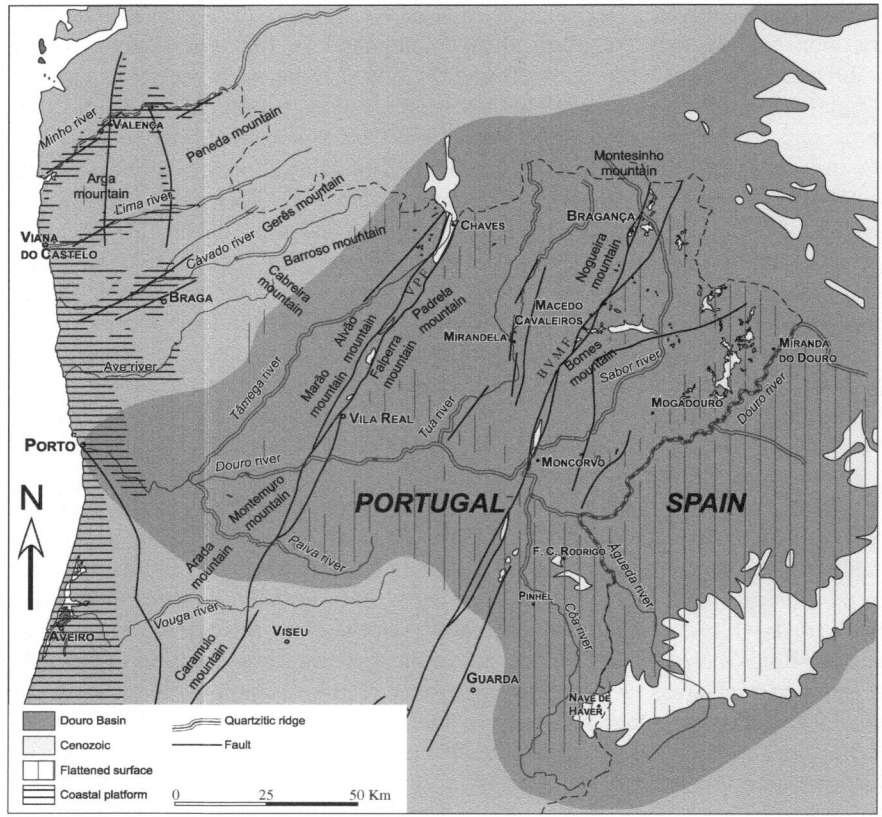

Fig. 3 Cenozoic deposits in NE Portugal

3 Douro Basin (Western Proximal Sector)

3.1 General Aspects

The Cenozoic Douro Basin (or Duero Basin, in Spanish) is the largest Cenozoic basin in the Iberian Peninsula, occupying a major part of the north-western region, and lying within the present-day Douro hydrographic basin (Fig. 1b). Only the western limit of the basin crops out in Portugal, namely in the Nave de Haver area. The other occurrences of the Douro Basin in Portugal referred to in this chapter are isolated from the main Spanish basin and are related to the infilling of small strike-slip basins and paleovalleys incised into the pre-Mesozoic basement of the western border of the basin (Figs. 2 and 3).

The Douro Basin is asymmetric, with the Neogene-Paleogene successions ranging up to only 400 m thick in the west and south, but reaching thicknesses of over 3000 m along the maximum subsidence axis running parallel to the northern-eastern border (Armenteros et al. 2002). The sedimentary record includes alluvial fan, fluvial and lacustrine deposits that have been divided variously according to different authors and different sectors of the basin (e.g., Santisteban et al. 1996; Armenteros et al. 2002; Alonso-Gavilán et al. 2004). The following division is generally accepted (Santisteban et al. 1996):

(i)   An Upper Cretaceous to Paleocene pre-orogenic tectosedimentary complex, with siliciclastic, carbonate and evaporitic deposits;

(ii) an Eocene to Oligocene tectosedimentary complex, split into several units mainly siliciclastic in character;

(iii) and a Miocene to Holocene tectosedimentary complex, split into several units with siliciclastic, carbonate and evaporitic deposits.

Up to the 1990s, Portuguese sediments in the western border area of the Douro basin, as described in this chapter, had been systematically included in the so-called *raña* facies of Iberia, a gravelly deposit with significant coverage in the Douro Basin and also in the Iberian Massif. *Raña* refers to an alluvial fan model, chronologically placed near the Plio-Pleistocene limit. This episode was related to the transition from the endorheic drainage of the Cenozoic Iberian basins to the present Atlantic drainage. More detailed investigations of these sediments (Pereira 1997, 1998, 1999, 2006; Pereira et al. 2000) have revealed a fluvial network in the proximal sector of the Douro Basin, and diverse tectonosedimentary units as described below.

3.2 Paleogene

The *Vilariça Formation* (Figs. 2, 3, 9 and 20) is described in the Vilariça (Torre de Moncorvo) and Longroiva (Mêda) strike-slip basins as a whitish or greyish-green heterometric sandy-gravel unit (Ferreira 1971, 1978; Cunha and Pereira 2000). The unit is moderately consolidated and has a quartz-feldspathic composition. These characteristics are similar to the arkose occurring at Nave de Haver (Almeida), along the south-western edge of the Douro Basin. On the Spanish side, this unit is refered as the Alamedilla Formation, indicated as Oligocene according to palynological data (Polo et al. 1987; Alonso-Gavilán et al. 2004). Correlation with Côja arkoses in the Mondego Basin supports the previous identification as Paleogene (Antunes 1964; Vallin 1965; Pereira 1997; Cunha and Pereira 2000).

In the Vilariça Basin, this unit is reasonably homogeneous and outcrops in a layer exceeding ~6 m in thickness at Cabanas de Cima (Torre de Moncorvo), in the southern sector of the depression. Sediments are mostly coarse sands with feldspar pebbles in a white muddy matrix with iron concretions. Some beds with planar structures define alluvial mantles. A 2 m thick layer of quartz-feldspathic gravels overlaps the sandy layer (Cunha and Pereira 2000). The unit has a tectonic tilt of 22° eastward (Pereira 1997). All features reflect a source in the porphyritic granite located to the west of the depression.

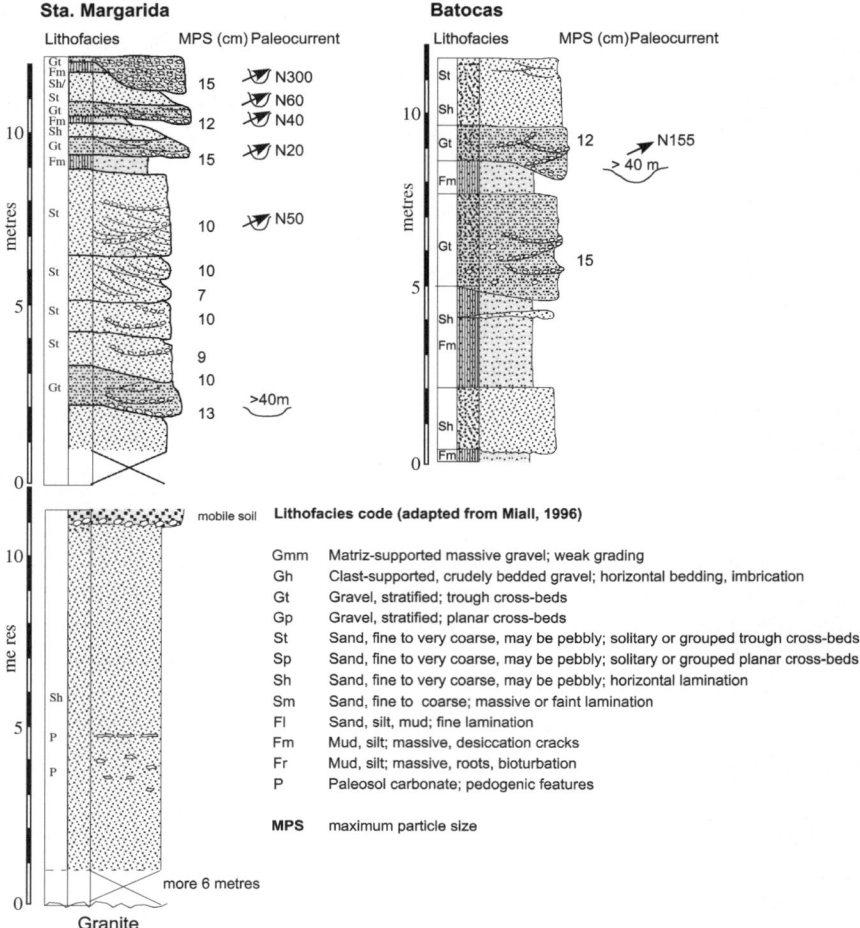

Fig. 4 Nave de Haver lithological logs

In the Longroiva basin (Mêda), ~7 m thick layer of sands (Longroiva fault and EN102 outcrops) are observed, passing upwards to 6-m thick alternating gravels and coarse to fine arkoses, representing large channel sequences. Quartz, granite and pegmatite pebbles have various shapes and reach 22 cm in diameter. The K-feldspar pebbles are fresh to moderately weathered, whereas plagioclase grains are weathered almost to clay. In the clay fraction, smectite and kaolinite are dominant over illite. Carbonate crusts are concentrated mainly in sub-vertical fractures (Pereira 1997; Cunha and Pereira 2000).

At Nave de Haver (Almeida) (Figs. 3 and 4), two members can be recognised in the outcrops of St. Margarida sand pit. The lower member consists of a ~50 cm thick succession of poorly calibrated sand-muddy layers. Small pebbles of quartz,

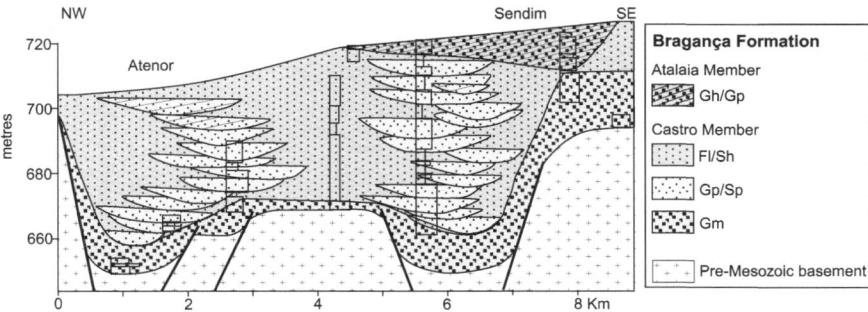

Fig. 5 Infilling model for the Bragança Formation in the confluence of the two major palevalleys of Atenor e Sendim in the Miranda do Douro Plateau area, with well developed flood facies

feldspar, and rare quartzite are supported in a greyish green sandy-mud matrix. This unit has reddish horizons and contains phreatic dolomite. The upper member, also cropping out at Batocas, is coarser, and exhibits wide fluvial channels. In general, sequences of channel-filling consist of gravel lithofacies followed by sand and, occasionally, muddy layers. Gravel composition reflects sources more distant than the surrounding granitic massif.

Considering data obtained from the Vilariça and Longroiva basins and Nave de Haver region, the Vilariça Formation consists of two members. The lower one is organized in planar layers of sandstones and mudstones deposited by alluvial mantles. The upper member is dominated by sequences of gravels and coarse to fine sands filling large channels, indicating an increase in the competence of the river and consequent channelling flows (Cunha and Pereira 2000).

It is assumed that the Vilariça Formation reflects, in the proximal sector, the inefficient drainage to the Douro Basin. The alluvial mantles were supplied by a granite source and developed on low-gradient surfaces, whose exhumation is represented by the Iberian Meseta surface (Ferreira 1971, 1978; Cunha and Pereira 2000). In the Vilariça and Longroiva basins, this unit can display pronounced tectonic tilting, and in the borders there is an over-thrust of the Variscan bedrock through faults with both reverse and horizontal components (Cunha and Pereira 2000). Fault characteristics are compatible with very intense regional Tortonian compression (Betic episode), and are responsible for the conservation of the Vilariça Formation in depressions.

3.3 Upper Miocene to Lower Pliocene

The *Bragança Formation* (Figs. 5, 6, 7, 8, 9 and 27) is defined in north-eastern Portugal as a lithostratigraphic unit recording a proximal fluvial paleodrainage to the Cenozoic Douro Basin in Spain (Pereira 1997, 1998, 1999; Pereira et al. 2000).

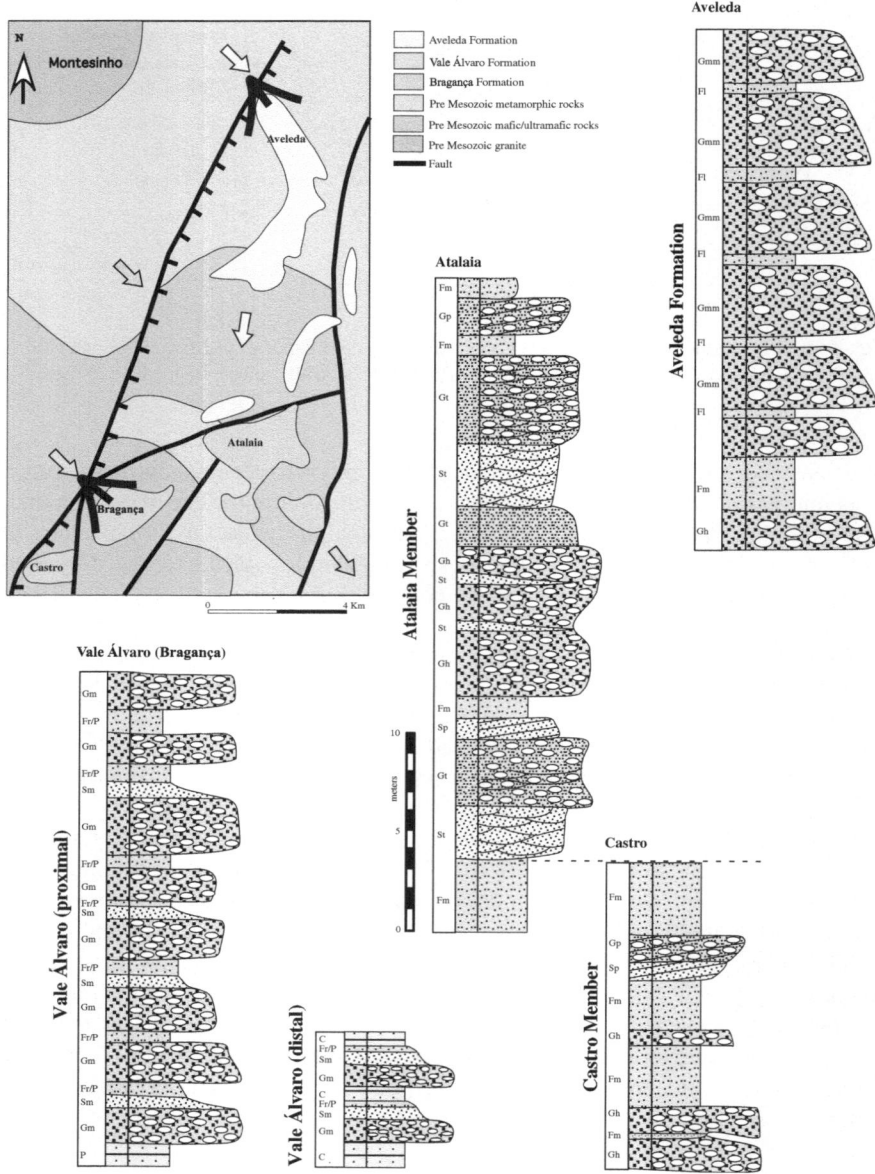

Fig. 6 Logs of the Bragança and Vale Álvaro Formations

It is recognized as being essentially Miocene, and may correspond to allostratigraphic units UBS11 and UBS12 (Fig. 9) identified in the Mondego Basin and in the proximal sector of the Lower Tejo Basin (Pereira 1997, 2006).

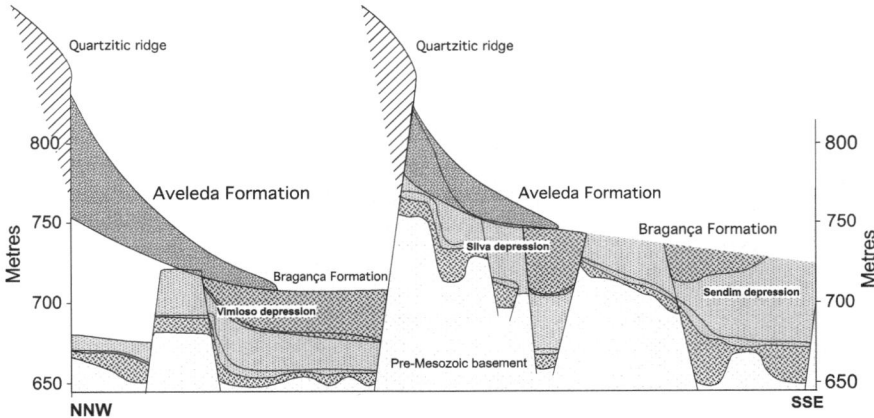

Fig. 7 Interpretational model of the tectono-sedimentary architecture of the Miranda do Douro plateau region. The filling of Vimioso-Silva-Sendim paleovalley by the sediments of the Bragança Formation is affected by tectonic movement along faults; subsequent sedimentation of the Aveleda Formation is associated with alluvial fans fed into the residual quartzite reliefs

The fluvial sediments fill incised paleovalleys incised in response to two tectonic episodes and the consequent orogenic uplift. These paleovalleys remain well preserved in Macedo de Cavaleiros (Vale da Porca, Salcelas and Castro Roupal outcrops), where they are oriented E–W, and to the east, near Vimioso (Vimioso outcrop) and Miranda do Douro (Silva, Sendim, and Atenor outcrops), where they are oriented NW–SE and NS, respectively. In this most north-eastern sector of Portugal, geophysical investigations and drilling (Moreira and Lima 1987) highlight the tectonic control of the paleovalleys of Sendim and Atenor, which have no morphological expression (Fig. 7). Given the context of the major Bragança-Vilariça-Manteigas (BVM) fault, the fluvial system was strongly influenced by the development of strike-slip basins (De Vicente et al. 2011).

The Bragança Formation comprises two members with similar compositions and depositional architectures. An unconformity is defined based on the recognition of a regional erosion surface and a well-developed paleosol. Deep channel gravel deposits and gravelly bars are the most characteristic lithofacies of the Bragança Formation. The sediments are immature, and contain moderately weathered feldspars in the sand fraction and a predominance of smectite and kaolinite in the clay fraction. The gravel units are mostly red in colour with lutite being brown, grey or green.

The Castro Member (lower member) has a thickness of about 70 m. In the most proximal area, especially in the tectonic strike-slip basins, the sedimentary infill is characterized by an increased thickness in massive gravelly units and by sandy channels. In these areas, the lithofacies and sedimentary architecture are compatible with the presence of alluvial fans that would have supplied the regional river system. Toward the top of the member, the sedimentary cycle is characterized by a thickening of fine sediment layers. This architecture is well represented

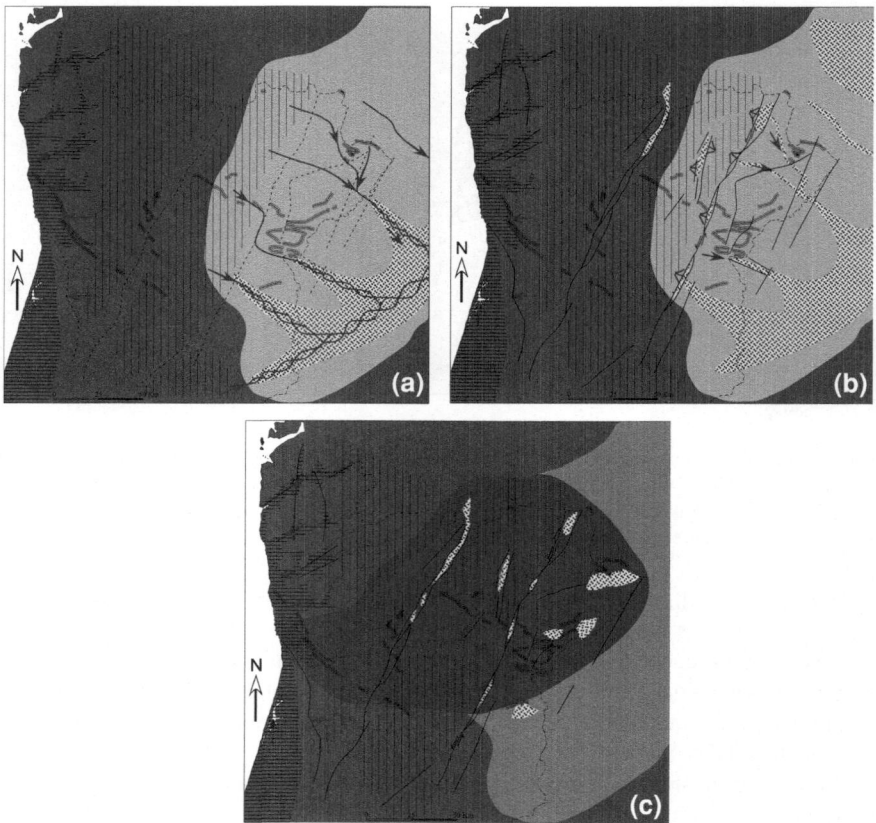

Fig. 8 Model of the paleogeographic stages of the western sector of the Douro Basin as documented by the sedimentary record. **a** Paleogene-Miocene: The drainage flows in two opposite directions, from relief raised in the central sector. It is assumed that there was a subdued geomorphological expression of the major faults. **b** Late Miocene: The infill of the Cenozoic Douro Basin and gentle erosion of the relief, with the definition of the fundamental surface of the Meseta, promotes progressive sedimentary filling of river valleys fed by alluvial fans generated by the reactivation of major NNE–SSW faults. Atlantic drainage gradually penetrates the high relief of the central sector. **c** Pliocene: Atlantic drainage captures the western sector of the Cenozoic Douro Basin. A pre-Douro river extends into the Sabor Basin. The basin of the Douro is pre-oriented with three main branches corresponding to the existing rivers Tâmega, Tua and Sabor. A new tectonic phase (Ibero-Manchega) reactivates the tectonic scarps and alluvial fans. Due to the radical change in the drainage, the more proximal areas in the periphery of the basin accumulate sediment at the base of the quartzitic reliefs. In contrast to the Sabor basin, which intensely dissects the Meseta surface, the Côa basin to the south preserves both the Meseta surface and the subsequent capture by Atlantic drainage

in the various sand pits of Sendas and Macedo de Cavaleiros regions and has a regional extent.

The basal unit of the Castro Member is best exposed in the area of Miranda do Douro Plateau, especially in the paleovalleys of Vimioso, Silva-Sendim and

Atenor (Fig. 5). It consists of channel-lag deposits of quartz and quartzitic clast-suported gravel, generated as the first response to tectonic uplift. Above, a generally thick lutite unit eroded by gravelly and sandy channels is identified, deposited during multiple infilling cycles.

The sedimentary characteristics of the Castro Member suggest a fluvial system of gravelly, braided rivers with low sinuosity, incised into the substrate bed (Miall 1996). The regional drainage framework and paleocurrent data show paleovalley river segments oriented mainly towards the south and east and a general direction of the river system towards the south-east.

The Atalaia Member (upper member) is less strongly expressed on the Miranda do Douro plateau, but reaches a thickness of about 50 m in the strike-slip basins of Bragança and Macedo de Cavaleiros and corresponds to a succession of episodes with similar characteristics to the Castro Member. The same fluvial environment, braided, low sinuosity rivers—resulted from a new tectono-sedimentary cycle. There is often a succession of the architectural elements of gravelly bars, derived from energetic fluvial episodes as well as lateral accretion elements, whose persistence indicates the temporary transition to a more sinuous river (Miall 1996). Occasionally, flood deposits are cut by abandoned channels.

The Bragança Formation is the most extensively exposed unit in north-eastern Portugal. Its particular characteristics include deposits reflecting different channel conditions and flows, the more organized facies, the weathering of clasts, the clay mineral contents with abundant smectite and neoformation of kaolinite, the development of incipient paleosoils, the red colour, and carbonate precipitation in incipient fractures. These characteristics suggest temperate to warm climatic conditions with a rainy season forming high energy streams, and a dry season that also favoured oscillation of the water table (Pereira 1997, 1998, 1999).

Geomorphological features suggest that, in the stage prior to the deposition of the Bragança Formation, river valleys incised in the bedrock developed as an erosional response to mountain uplift occuring since the Eocene (Pereira et al. 2000; De Vicente et al. 2008, 2011). Subsequently, the activity of this major tectonic episode, which corresponds to the Betic compression at about 9.5 Ma (Calvo et al. 1993), caused a staircase organization of large fault blocks in northern Portugal and the development of strike-slip basins associated with the BVM fault detachment (Ferreira 1991; Cunha 1992; Cabral 1995). It is most likely that the Castro Member represents the response of the alluvial systems to the new morphotectonic regime (Pereira et al. 2000).

The *Vale Álvaro Formation* occurs in the eastern region of Portugal (Fig. 6) unconformably overlying the mafic rocks that constitute the Bragança Massif (Vale Álvaro and IP4 road outcrops, in Bragança city) and the Morais Massif (Limãos outcrops near Macedo de Cavaleiros city). It has traditionally been indicated as probably Paleogene, based on sedimentological characteristics (Ramalhal 1968; Pereira and Azevêdo 1991; Pereira 1997, 1998). More recently, palynological data (Poças et al. 2003; Poças 2004) have suggested an age between the Late Miocene and Pliocene, and a climate characterized by seasonal contrasts (wet winters and hot summers) and overall moderate temperatures.

A sedimentary characterization of the Vale Álvaro Formation shows several episodes of coarse sediments consistent with an alluvial fan model. Pedogenic and phreathic calcretes, widespread in the distal area, developed between the debris flow episodes.

The proximal lithofacies of the Vale Álvaro Formation are exposed along the railway line in Bragança city. The lower part consists of a calcrete layer, followed by several successive basic sequences, beginning with red, coarse, clast-supported gravel. Sub-rounded clasts are exclusively of weathered mafic and ultramafic rocks from the Bragança Massif. The larger clasts reach 60 cm in diameter. Sedimentary basic sequences include sand and marl layers. The sand fraction consists mainly of fragments of mafic and ultramafic rocks and grains of hornblende and garnet. The clay fraction is composed mainly of dioctahedral montmorillonite and paligorskite, chlorite, interstratified chlorite-montmorillonite and occasionally serpentine, talc and kaolinite (Pereira and Brilha 2000). The entire unit is consolidated by ferruginous and carbonate cement (Pereira 2006).

Currently, outcrops of the Vale Álvaro Formation are limited to the distal sector of the alluvial fan, seen along the IP4 road and its entry to the city of Bragança. With increasing distance from Bragança, the unit thins, particle dimensions are reduced, and calcrete layers developed on paleosoils increase (Pereira 2006). In Limãos (Macedo de Cavaleiros), there is a further development of a 4-m thick dolocrete that replaces the amphibole shale bedrock. Outcrops of sediments are limited to a 3-m thick gravel layer with clasts of mafic rocks underlying the Bragança Formation (Pereira 2006).

The morphotectonic context, sedimentary characteristics and depositional architecture of the Vale Álvaro Formation indicate an alluvial fan model, which relates to an episode of tectonic movement that stressed the fault scarps (Pereira 1997, 2006). The Bragança alluvial fan is clearly of tectonic origin, associated with the BVM fault.

The age of the Vale Álvaro Formation remains unclear, but it might be a proximal equivalent of the Bragança Formation with epigenetic carbonate developed along the contact with basic rocks lying in tectonic depressions.

3.4 Upper Pliocene

The *Mirandela Formation* (Figs. 9, 27), with a thickness exceeding 30 m in Mirandela basin (Vila Nova das Patas and Eixes outcrops), consists of whitish or yellowish sediments. The unit is formed by a succession of sequences of clast-supported gravelly layers, sandy layers and, more rarely, muddy layers. The clasts, moderately rounded, are primarily of quartz, quartzite, and kaolinite dominates the clay fraction (Pereira 1997).

The sedimentological features suggest a fluvial model of high energy and open drainage, filling paleovalleys. It is assumed that these conditions may have been reached during the Late Pliocene, with the opening of the Mirandela depression to

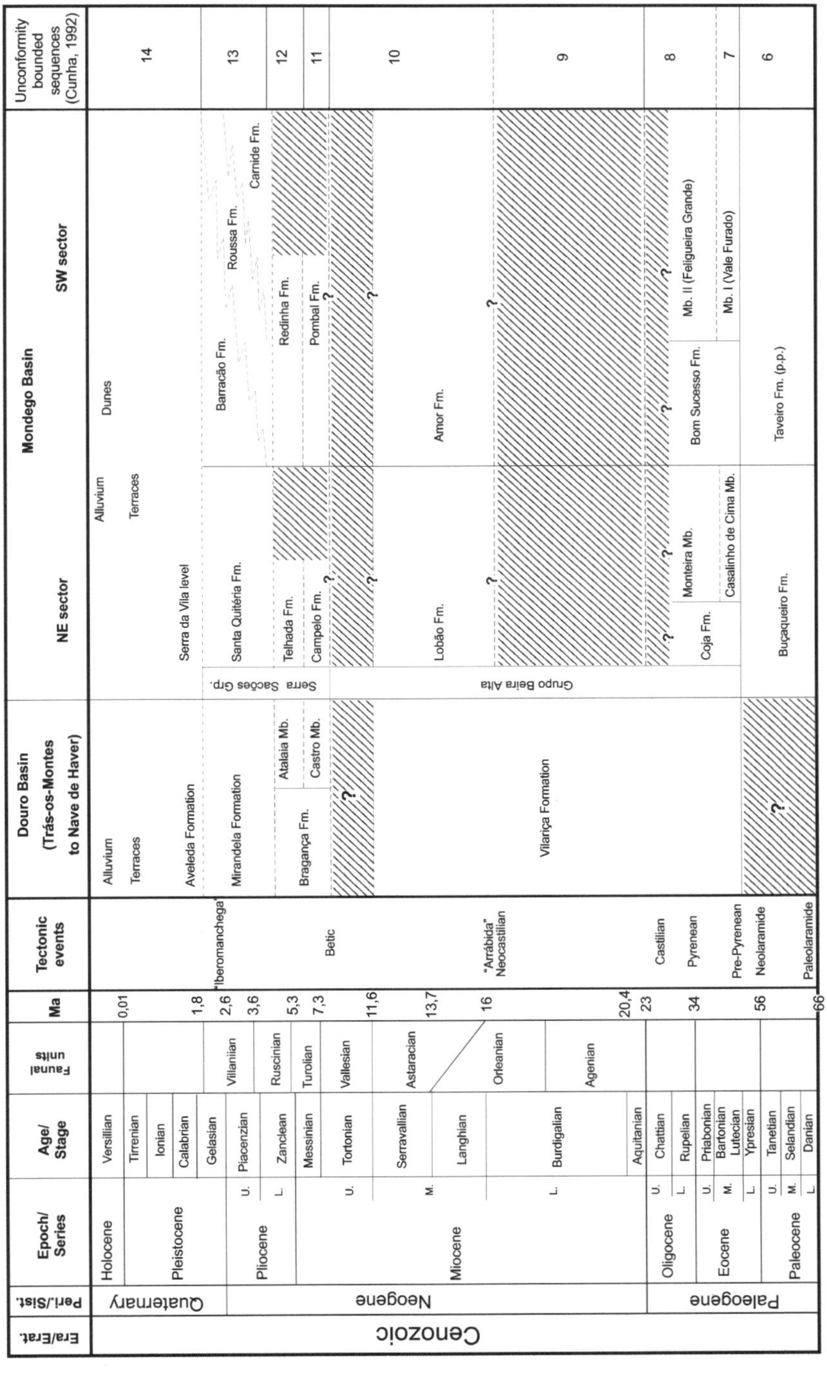

Era/Erat.	Peri./Sist.	Epoch/Series	Age/Stage	Faunal units	Ma	Tectonic events	Douro Basin (Trás-os-Montes to Nave de Haver)	NE sector	Mondego Basin SW sector	Unconformity bounded sequences (Cunha, 1992)
Cenozoic	Quaternary	Holocene	Versilian		0,01		Alluvium	Alluvium	Dunes	14
		Pleistocene	Tirrenian / Ionian / Calabrian		1,8		Terraces / Aveleda Formation	Terraces / Serra da Vila level		
			Gelasian		2,6	"Iberomanchega"	Mirandela Formation	Santa Quitéria Fm.	Barracão Fm. / Roussa Fm. / Carnide Fm.	13
	Neogene	u. Pliocene	Piacenzian	Villanian	3,6		Bragança Fm. / Atalaia Mb. / Castro Mb.	Telhada Fm. / Campelo Fm.	Redinha Fm. / Pombal Fm.	12
		l. Pliocene	Zanclean	Ruscinian	5,3	Betic				11
		u. Miocene	Messinian	Turolian	7,3		Vilariça Formation	Lobão Fm.	Amor Fm.	10
			Tortonian	Vallesian	11,6					
		M. Miocene	Serravallian	Astaracian	13,7	"Arrábida" Neocastilian				9
			Langhian	Orleanian	16					
		l. Miocene	Burdigalian	Agenian	20,4	Castilian		Coja Fm. / Monteira Mb. / Casalinho de Cima Mb.	Bom Sucesso Fm. / Mb. II (Feligueira Grande)	8
			Aquitanian		23	Pyrenean			Mb. I (Vale Furado)	7
	Paleogene	u. Oligocene	Chattian			Pre-Pyrenean		Buçaqueiro Fm.	Taveiro Fm. (p.p.)	6
		l. Oligocene	Rupelian		34					
		M. Eocene	Priabonian / Bartonian / Lutecian			Neolaramide				
		l. Eocene	Ypresian		56					
		u. Paleocene	Tanetian			Paleolaramide				
		M. Paleocene	Selandian							
		l. Paleocene	Danian		66					

Serra Sacões Grp. / Grupo Beira Alta

Fig. 9 Correlation between lithostratigraphic units from the Cenozoic Douro (Trás-os-Montes to Nave de Haver) and Mondego Basins and the unconformity-bounded sequences (UBS's) (updated, Cunha et al. 2009)

Atlantic drainage, probably involving a precursor of the Douro river (Fig. 8c) (Pereira 1997, 1998, 1999; Pereira et al. 2000).

3.5 Lower Pleistocene

The *Aveleda Formation* (Fig. 9) consists of reddish deposits mainly of muddy matrix-supported gravel. Clasts of several metasedimentary rocks and quartz are subangular. Kaolinite and illite dominate the clay fraction.

The Aveleda Formation is generally located in two different geomorphological settings. First, occurrences such as Vale de Frades (Vimioso), Sra do Nazo (Miranda do Douro), Souto da Velha and Carviçais (Torre de Moncorvo) lie over a flattened surface, which marks a discontinuity with the older units, and have it source on the resistant reliefs. Second, occasionally this unit lies within tectonic depressions associated with the BVM fault. The most important outcrops in this context are those in: Aveleda (north of Bragança), with Montesinho Mountain as a source; Sortes (south of Bragança), Azibeiro, Podence and Arrifana (Macedo de Cavaleiros), with Nogueira Mountain as a source; and Castelãos (Macedo de Cavaleiros), with Bornes Mountain as a source (Pereira 1997).

The lithofacies and architecture of Aveleda Formation indicate nearby sources and debris flows deposited as alluvial fan bodies. This unit establishes the transition between the previous endorheic drainage network and the Atlantic fluvial network (Pereira 2006) (Fig. 7).

The Aveleda Formation correlates with other sedimentary units, the so-called Iberian *raña* facies, described in the Iberian Peninsula, which follow, in general, the infilling of the Cenozoic Iberian basins and subsequent development of an important erosional episode represented by the most continuous surface of the Iberian Meseta (Aguirre 1997). Given the presence of the formation inside strike-slip basins associated with the BVM fault, it is assumed to be related to the 2.0 Ma tectonic episode identified in other Iberian basins (Calvo et al. 1993; Pereira 1997, 2006).

4 Mondego Basin

4.1 General Aspects

The general trend of the Lower Cretaceous of the Portuguese western margin reflects the transition from late rifting to passive margin, with the last break-up unconformity dated as late Aptian (Dinis et al. 2008). The western margin of Iberia, the post-rift main allostratigraphic unit (Unconformity Bounded Sequence UBS 5; Cunha 1992), reaching ~300 m in thickness and probably comprising the upper Aptian to the lower Campanian, was deposited (Cunha and Reis 1995). This unit covered a vast planation surface and was fed by a source area located to the NE, comprising mainly Variscan granites. The sedimentary record is dominated by siliciclastics and comprises fluvial and deltaic coastal marine siliciclastic systems, as well as extensive deposits of a shallow marine carbonate platform. The Lusitanian Basin was completely infilled by the late Cenomanian, when a large carbonate platform extended far inland. With very low accommodation volumes being created throughout the Late Cretaceous, fluvial deposits were dominant, but a few marine layers are recorded related to eustatic rises that occurred in the early Turonian, Coniacian, early Campanian and Maastrichtian.

The increasing N–S Pyrenean compression led to the cessation of the opening of the Gulf of Biscay and to an increase in transpressive intraplate stress. This caused important palaeogeographic changes on the Portuguese mainland, comprising volcanism, diapirism along N–S-trending faults and vertical displacement along NE–SW-trending faults, probably with a left-lateral transpressive component. One of the active faults was the Lousã Fault, whose southern block was uplifted leading to a second cycle of sedimentation generated by the erosion of previous sediments (Buçaco Group).

4.2 Upper Campanian to Lower Eocene

The sediments of UBS 5 (Upper Cretaceous) overlie a thick regional silcrete at the top of UBS 4, but along diapirs they cover an angular unconformity. The silcrete indicates long-term landscape stability and subtropical weathering. Onshore, the uppermost Campanian to Lower Eocene succession, with a maximum thickness of ∼ 200 m, consists of an alternation of yellow sands and red to green silts with nodular calcareous crusts, representing a coastal plain environment with meandering fluvial systems draining to the NW (*Buçaqueiro* and *Taveiro Formations*, Fig. 9); these sediments change distally into lagoonal and restricted shallow marine deposits, with barrier-island facies in the Aveiro region (*Aveiro Formation*; Bernardes and Corrochano 1987).

The Taveiro and Aveiro formations have provided fossils ascribed to the late Campanian and Maastrichtian (Antunes and Pais 1978; Antunes 1979a; Reis 1981). Their offshore equivalents are the very shallow marine fine clastics and dolomites of the upper Campanian–Maastrichtian Dourada Formation (90 m thick). This unit consists of grey-brown, dolomite-cemented quartzarenites grading to sandy crystalline dolostones with intercalations of grey to light brown marls and occasionally sandy limestones in the lower part (Witt 1977). Correlative N–S diapiric reactivations built up coalescent alluvial fans, composed of conglomerates with limestone clasts and red silts (Reis 1983). The succession displays a general fining-upward trend, although the architecture can be rather complex.

The UBS 6 unit is represented by the *Silveirinha Formation* (Fig. 9), which comprises an alternation of lenticular beds of calcitic conglomerates with abundant fossil remains, laminated and cross-bedded sands, and brownish-red silts (Reis 1981; Reis et al. 1981; Soares and Reis 1982; Soares et al. 1986).

The Silveirinha site (south of Figueira da Foz) has provided a very rich vertebrate faunal assemblage, including *Peratherium* cf. *constans, Apatemys* div. sp., *Heterohyus* div. sp., *Russellmys denisae, Leptacodon nascimentoi, Didelphodus* sp., *Arcius zbyszewskii, Eurodon silveirinhensis, Donrussellia lusitanica, Miacis* cf. *latouri, Microparamys paisi, M.* cf. *nanus, Meldimys cardosoi, Paschatherium marinae, Microhyus reisi, Diacodexis antunesi, Hyracotherium* cf. *vulpiceps,* and *Illerdoryctes* cf. *sigei* of the Thanetian-Ypresian (the transition from Paleocene to Eocene) (Antunes in Reis et al. 1981; Antunes and Russel, 1981; Antunes et al. 1987b; Estravis 1992, 2000; Antunes et al. 1997a). More recently, the Charophytes, molluscs, ostracoda, fish, amphibians and reptiles have also been studied (Antunes and Colin 2003; Callapez 2003; Colin and Antunes 2003; Antunes and Gaudant 2003; Rage and Augé 2003; de Broin 2003).

4.3 Middle Eocene to Oligocene

By Middle Eocene times, intense compression at the Pyrenees induced lithosphere folding of Iberia. To the west, the Cenozoic Mondego Basin started to open in a SW–NE direction. The basal infill of the basin is Middle Eocene to Oligocene (?) in age and comprises mainly fluvial arkoses. The main lithostratigraphic units are represented by the Bom Sucesso Formation and the Coja Formation (the proximal equivalent).

The *Bom Sucesso Formation* (Figs. 9 and 20) includes gravels at the base, passing upwards to sands and silts (Reis 1981, 1983). It is represented in the SW of Coimbra and has an average thickness of 50–60 m, locally reaching 120 m. Member I of the formation is 70 m thick and is well exposed in the Vale Furado and Feligueira Grande coastal clifs. The deposits consist of reddish gravels, sands and silts that provide fauna including *Iberosuchus macrodon, Paralophiodon* cf. *leptorhynchum,* turtles and small mammals (not identified). Somewhat higher layers contain *Paranchilophus lusitanicus* (Antunes 1975, 1986a, b, c, 1995). The deposits of the lower member disconformably contact the yellow coarse sands of member II, 40 m thick and dated Late Eocene. The fossils found imply ages of Middle Eocene to Late Eocene for layers in member I and the end of the Late Eocene for the base of member II.

The *Coja Formation* (Cunha et al. 1992; Cunha 1991, 1992, 1999a, b) (Figs. 9 and 20) comprises the sedimentary record of the sector to the NE of Coimbra. Here, the unit overlays a vast planation surface cut on Cretaceous sediments, metamorphic rocks of the Douro Group and Variscan granitoids. However, hard quartzite ridges located in Ordovician synclines constituted paleoreliefs transverse to the exorheic drainage (Cunha and Reis 1995). The sedimentation in the elongated SE-SW alluvial plain was controlled by small, sub-vertical NNE–SSW faults (Cunha 1992).

The Coja Formation corresponds to a sandy succession deposited in the upstream sector of a braided alluvial plain, with drainage generally directed to the SW and SSW towards the Atlantic, crossing the westernmost extremity of the yet-to-be-uplifted Portuguese Central Range. The formation consists mainly of coarse arkoses, sometimes gravelly, and has been measured up to 73 m in thickness. The cobbles are composed of quartzite, white quartz, phylite/metagreywacke and granite. The sands, greyish green or whitish, sometimes with purple spots, are rich in feldspar, but also contain mica, tourmaline and rutile. The clay minerals comprise smectite (abundant), illite and kaolinite. Some basal layers exhibit dolomite cementation and paligorskite, resulting from a phreatic origin (eodiagenesis) or related to pedogenic processes (Cunha 2000).

The upper limit of the formation corresponds to a regional unconformity. The Coja Formation includes two members, separated by a sedimentary discontinuity (Antunes 1967; Cunha 1999a). The Casalinho de Cima Member (the lower) is up to 43 m thick and consists of coarse massive arkoses. It can contain lenses of grey silt with plants (spores, pollen - mosses, liverworts, licophytes, pteridophytes, and angiosperms, conifer, and incarbonized trunks of a fern and a meliaceae or a leguminosae; Pais 1992). The Monteira Member, up to 30 m thick, has basal gravelly facies, but sands and silts from there to the top. At Coja, and further north in Naia, silty layers have provided fauna (*Geochelone* sp., *Peratherium* cf. *cuvieri*, *Palaeotherium* cf. *crassum*, *Palaeotherium magnum*, *Diplobune secundaria*, and *Anoplotherium* cf. *commune*) ascribed to the Ludian (Uppermost Eocene; Antunes 1964, 1967, 1986a, b, c; Antunes and Broin 1977; Antunes et al. 1997a). At Sobreda (near Carregal do Sal), a silicified trunk of *Cupressinoxylon lusitanensis* (*Tetraclinis* sp.?) has been discovered (Pais 1992), also contained in the Nave Haver arkoses (Vallin 1965).

The sedimentological and palaeontological data indicate that, during the deposition of the Casalinho de Cima Member, the vegetation was rich, with forests and wet conditions; the climate was subtropical (Pais 1992). The characteristics of the Monteira Member indicate that the climate was probably temperate and dry.

The Coja Formation is the lateral equivalent of the Bom Sucesso Formation, with a paleogeographic continuity in the Mondego Basin. Both formations are composed of two members, separated by a disconformity that distinguishes allostratigraphic units UBS 7 (Middle to Upper Eocene) and UBS 8 (Upper Eocene and part of the Oligocene ?). Thus, the Casalinho de Cima and Monteira members correlate, respectively, with members I and II of the Bom Sucesso Formation. Based on the fossil localities with chronostratigraphic significance (Vale Furado, Coja and Naia), the lower member corresponds to the Middle Eocene and the Upper Eocene, while the upper member corresponds to the Uppermost Eocene and possibly also the lower Chatian (Reis and Cunha 1988, 1989).

4.4 Middle Miocene

The Mondego basin includes a Miocene stage of infilling, the sedimentary record of which is represented by allostratigraphic unit UBS 10, and the lithostratigraphy comprises the Amor Formation and its proximal equivalent (NE of Coimbra) the Lobão Formation.

The *Amor Formation* (Reis 1983; Cunha and Reis 1989) (Figs. 9 and 23) records a braided alluvial plain developed with poor but exorheic drainage to the Atlantic Ocean (to the SW). It includes an alternation of thick beds of sands and silts with an assembled fauna of gastropods, fish, amphibians, reptiles, birds, and mammals, including carnivores, artiodactyls, perissodactyls, proboscideans, insectivores, lagomorphs, rodents including *Hispanotherium matritense* (also known in Div. Vb Lisboa, of the Lower Tejo Basin) attributed to the Aragonian (Mammal Zone MN5, equivalent to the lower Langhian), and a new species of cricetidae, *Fahlbuschia freudenthal* (Zbyszewski and Ferreira 1967; Antunes and Mein 1981). The host fauna point to a marshy environment, crossed by streams. The prevailing climate was warm and relatively dry (Antunes and Mein 1981).

The *Lobão Formation* (Figs. 9 and 23), defined in the area NE of Coimbra, consists of coarse green-orange, poorly sorted arkoses. The formation is almost exclusively sandy, with minor lenses of silt. The larger clasts (<12 cm) are of quartzite, white quartz, feldspar and granite. Channel geometries reaching 100 m in width are typical. Layers with bioturbation, containing traces of fossil roots, can be identified. Although laterally extensive, the formation is only 4 m thick. The present remains of the unit are restricted to the area between Tondela and Óvoa.

The unit overlies the Coja Formation or, towards the NE, granite rocks. Given its distance from the piedmonts of the Caramulo and Estrela mountains, the unit was not covered by alluvial fan deposits related to the Late Miocene to Pliocene uplift and erosion of these relieves (and deposition of the de Sacões Group). The upper limit of the Lobão Formation represents the culminant level of sedimentary filling in this restricted area, and appears to correspond to the last stage of elaboration of the vast planation surface cut into the Beiras granites, degraded by the incision of the Quaternary fluvial network.

The deposits of the Lobão Formation correspond to the proximal sector of a braided alluvial plain, draining towards the SW. Compared with the Coja Formation, the Lobão Formation contains coarser sediments with less matrix, and is richer in feldspars, which are less weathered and larger. Overall, the depositional facies reflect a better organized drainage. The clay minerals include kaolinite (abundant) and illite, in contrast to the Coja Formation, which is rich in smectite. It is likely that a hot, dry climate and poor drainage characterized the depositional environment of the Lobão Formation.

4.5 Upper Miocene to Lower Pleistocene

The *Serra de Sacões Group* (Figs. 9 and 26) is ascribed to the upper Tortonian to Gelasian (Cunha 1999), or even to the earliest Pleistocene. The group includes alluvial fans deposited along the Lousã and Verín-Penacova fault scarps. The deposits are heterometric, consisting of gravels with intercalations of silts. The group overlies the Paleogene (Coja Formation), the Cretaceous (Buçaco Group and the Buçaqueiro Sands) or the Hercynian basement. The top of the Serra de

Sacões Group represents the culminant level of the sedimentary infilling. The fluvial incision, which probably started after the end of the Gelasian, was initiated by the development of the Serra da Vila layer (100 m below the top surface of the group; Daveau et al. 1985–86).

The Serra de Sacões Group includes the allostratigraphic units UBS 11, UBS 12 and UBS 13. It represents the sedimentary response to tectonic movements during the Late Miocene and Pliocene that led to the differentiation of the present relief (Daveau 1985; Ribeiro et al. 1990a; Ferreira 1991; Cunha 1992; Cunha and Reis 1992; Sequeira et al. 1997), such as the Estrela and Caramulo mountain systems. With increasing distance from the source area, the alluvial system decreases in thickness and grain size, from gravels to sandy silts.

The Serra de Sacões Group includes, from the base to the top, the Campelo, Telhada and Santa Quitéria formations (Cunha 1999a). Along the Leiria-Pombal fault scarp and in the Ourém area, similar alluvial deposits are represented.

The *Campelo Formation* (UBS 11, upper Tortonian to Messinian) (Figs. 9 and 26) consists of a fining-upwards sequence, with a predominance of gravels at the base and silts towards the top. The unit has abundant clasts of phyllite and meta-greywacke. The clay minerals include abundant illite, associated with kaolinite and smectite. The regional variation of facies allows two members to be distinguished. The gravels have a dominant green, orange or red color and can reach 100 m in thickness (Folques Member). The distal sectors show green or yellowish grey silts (Arroça Member). The Campelo Formation disconformably overlies the Paleogene (Coja Formation), the Cretaceous (the Buçaqueiro Formation and the Buçaco Group) or the Palaeozoic basement. The alluvial fan sedimentation was endorheic.

The *Telhada Formation* (UBS 12, upermost Messinian to Zanclean) (Figs. 9 and 26), reaches 90 m in thickness and unconformably overlies the Campelo Formation. It presents an intense red color, which results from a clay matrix containing iron oxides (mainly goethite): the iron oxides penetrate into the nucleus of the phyllite/metagreywake or milky quartz clasts. With increasing distance from the source area, the formation gradually decreases in particle size and in thickness, and the facies became more organized. Along the fault scarps that were active at the time, the formation is composed exclusively of matrix-supported gravels, but distally the overall thinning is associated with a transition to alternating gravels and silts. The clay minerals include illite and kaolinite, in similar quantities. A phase of uplift of the Portuguese Central Cordillera and other mountainous terrain triggered torrential sedimentation in the form of alluvial fans with endorheic drainage. A temperate climate, but with highly contrasting seasonal conditions and wide fluctuations in groundwater level, is inferred, based mainly on the appearance of rubefaction and the clay mineralogy (Cunha 2000).

The *Santa Quitéria Formation* (UBS 13, upermost Zanclean to Gelasian) (Figs. 9, 10 and 26) records the upper part of alluvial fan sedimentation. In the Sacões hill area it attains 250 m in thickness, and the surface at 600 m altitude (a.m.s.l.), represents the culminant surface of the sedimentary basin. At the foot of the Penedos de Góis quartzite ridges, the unit is heterometric, with blocks reaching 4 m in size and supported by a silty-sandy matrix. Towards the SSW (downstream), the deposits are finer and exhibit tractive sedimentary structures. The cobble and boulder clasts are mainly of quartzite, associated with some that are phyllite/ metagreywacke or white quartz. Clay minerals include kaolinite (predominant), illite and vermiculite. The unit is a brown yellowish color, sometimes whitish or reddish. It unconformably overlies the Zanclean (Telhada Formation), but also Cretaceous and Triassic sedimentary units.

Deposition of the Santa Quitéria Formation occurred in a fan delta draining towards the WNW into the Atlantic coast (Cunha et al. 1993). The deposits and the substrate at the unit base show features such as kaolinization and hydromorphism, which indicate weathering processes associated with significant leaching and well-drained conditions. Frequent rains and intense weathering of phyllites in the source area would likely have induced mass movements on the proximal slopes. The spatial development of the alluvial facies, the predominance of clasts highly resistant to weathering, the intense transformation of phylites into clay-rich materials, the predominance of kaolinite (Cunha 2000) and the particular plant fossils (Vieira 2009), all suggest the persistence of a hot and humid climate. Studies of marine fossils (Silva 2001) indicate a sea water temperature warmer than the present, probably tropical.

The deposition of this unit post-dates a compressive tectonic event but also predates a later one, as documented by the unit being thrusted (NE–SW-trending Lousã Fault) by the basement of the uplifted block (Portuguese Central Cordillera). The unit also pre-dates the beginning of the stage of fluvial incision, expressed by a succession of terraces and colluvium deposits (Ramos et al. 2009).

South-west of Coimbra, the progressive facies change revealed in the sedimentary record of the UBS 13 allostratigraphic unit (Fig. 10) leads to the identification of the *Barracão Group* (Cunha et al. 1992, 1993; Ramos and Cunha 2004) (Figs. 9, 10 and 27). This group includes a typical succession that includes, from the base to top: marine yellow sands, the *Carnide Formation*; white deltaic sands, the *Roussa Formation*; an alternation of 0.5–1 m thick beds of palustrine silt (sometimes organic and containing fossil trunks) and sand beds, grading upwards to fluvial gravels (*Santa Quitéria Formation*).

The lower limit of the Barracão Group is a disconformity or an angular unconformity on the Miocene (Amor Formation) or Mesozoic substratum. In the areas where the base of this unit consists of marine sediments, the sedimentary discontinuity corresponds to a transgressive surface expressed by an extensive wave-cut platform. The upper limit of the group is a vast sedimentary surface, generally varying from ~ 100 to ~ 60 m in altitude (a.m.s.l.; Ferreira 2005), and deformed by tectonic movements. This sedimentary surface became abandoned once the streams draining the coastal plain started to incise. The base of the Carnide Formation contains the NN 16 nannofossil biozone and has been dated at ~ 3.6 Ma (transition Zanclean-Piacenzian) (Cachão 1989, 1990; Silva 2001; Silva et al. 2010).

The Barracão Group has a tabular geometry, with its thickness increasing gradually towards the west. However, syn-depositional tectonics seem to have been responsible for local variations in both thickness (up to 40–70 m, instead of the usual 20 m) and facies. Usually, the cobble and boulder clasts are of quartzite (predominant) and quartz. Locally, the base of the marine unit includes blocks of

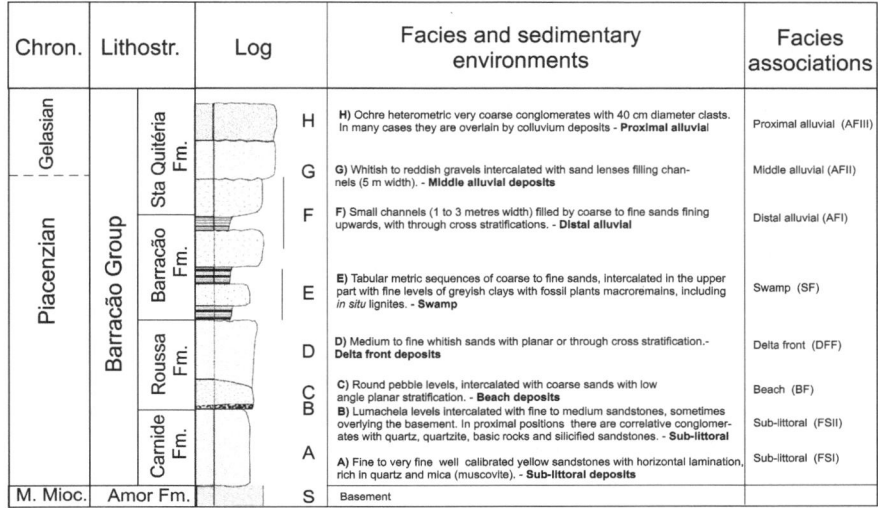

Chron.	Lithostr.	Log	Facies and sedimentary environments	Facies associations
	Sta Quitéria Fm.	H	**H)** Ochre heterometric very coarse conglomerates with 40 cm diameter clasts. In many cases they are overlain by colluvium deposits - **Proximal alluvial**	Proximal alluvial (AFIII)
Gelasian / Piacenzian / Barracão Group	Barracão Fm.	G	**G)** Whitish to reddish gravels intercalated with sand lenses filling channels (5 m width). - **Middle alluvial deposits**	Middle alluvial (AFII)
		F	**F)** Small channels (1 to 3 metres width) filled by coarse to fine sands fining upwards, with through cross stratifications. - **Distal alluvial**	Distal alluvial (AFI)
		E	**E)** Tabular metric sequences of coarse to fine sands, intercalated in the upper part with fine levels of greyish clays with fossil plants macroremains, including *in situ* lignites. - **Swamp**	Swamp (SF)
	Roussa Fm.	D	**D)** Medium to fine whitish sands with planar or through cross stratification.- **Delta front deposits**	Delta front (DFF)
	Carnide Fm.	C / B	**C)** Round pebble levels, intercalated with coarse sands with low angle planar stratification. - **Beach deposits**	Beach (BF)
			B) Lumachela levels intercalated with fine to medium sandstones, sometimes overlying the basement. In proximal positions there are correlative conglomerates with quartz, quartzite, basic rocks and silicified sandstones. - **Sub-littoral**	Sub-littoral (FSII)
		A	**A)** Fine to very fine well calibrated yellow sandstones with horizontal lamination, rich in quartz and mica (muscovite). - **Sub-littoral deposits**	Sub-littoral (FSI)
M. Mioc.	Amor Fm.	S	Basement	

Fig. 10 Lithological types of UBS 13 (modified from Ramos 2008)

silicified sandstone reaching 0.8 m in diameter, which probably represent the remains of coastal erosion of some of the more resistant sedimentary units, namely Cretaceous and Paleogene well-cemented sandstones (Ramos 2008). The sands are rich in quartz, with the feldspars being very weathered, and the clay minerals consist of predominant kaolinite and some illite.

A paleogeographic reconstruction for the Zanclean-Piacenzian transition ($\sim$3.6 Ma) (Fig. 26) depicts a penetrative marine transgression that produced a shallow sandy littoral zone, with an abundant siliciclastic supply sourced from the uplifted terrains to the east by fan-deltas corresponding to the ancestors of the Mondego, Arunca, Carnide and Lis rivers (Ramos and Cunha 2004; Ramos 2008).

It can be assumed that the main control for the important marine incursion and coeval accommodation space generated was the Pliocene eustatic maximum, estimated at 40–60 m a.m.s.l. (Dowsset et al. 1996). In the context of the high sea level, a rapid westwards progradation of continental siliciclastic systems occurred, and sourced the important siliciclast continental input (Fig. 26).

5 Lower Tejo Basin

5.1 General Aspects

The Lower Tejo Basin (LTB) occupies a large area (260 km long and 80 km wide) in Portugal, and extends from the coastal region of Lisboa-Setúbal Peninsula to beyond the border with Spain near Castelo Branco (Beira Baixa). Its genesis is related to the Pyrenean compression. The sedimentary infilling began in the Middle Eocene, while the transition to the extant fluvial incision occurred in the end of the Gelasian.

Three distinct sectors can be recognized:

1. The distal sector, in the south-western Lisboa and Setúbal Peninsula areas, in close proximity to the Atlantic;
2. the central sector, comprising the Ribatejo and Alto Alentejo, with continental facies and some brackish episodes corresponding to higher eustatic sea levels; and
3. the proximal sector, in southern Beira Baixa, crossing the Portugal-Spain border, containing continental facies only (Figs. 1 and 2).

Lisboa and Setúbal Peninsula are located in the distal sector of the basin, in which Miocene beds are well exposed. Miocene sedimentation patterns correspond to temporal changes in the extents of marine and continental domains.

Early studies of geological aspects of the Lower Tejo basin included those of Sharp (1834, 1841), Dollfus et al. (1903–04), Choffat (1950), Cotter (in Dollfus et al. 1903–04; 1956), and Zbyszewski (1941, 1954, 1957, 1962, 1963, 1964, 1967). However, as new concepts in sedimentary interpretation became established, including rythmic sedimentation, erosion and sedimentation cycles, sequences and corresponding types, new insights into the stratigraphy of the Lower Tejo basin were developed. A high-resolution stratigraphic strategy has been devised, comprising:

(a) Detailed studies of important sections (Antunes and Torquato 1969-1970, and subsequent investigations);
(b) The improvement of dating and other data by means of biochronologic research, mostly on mammals and planktic foraminifera, K–Ar and (later) Sr isotope ages (Table 1), sedimentology, paleoecology, and magnetostratigraphy, making possible the development of regional syntheses and broad-scale correlations (Antunes et al. 1973, 1987a, 1996a, b; Antunes and Pais 1984, 1992a, 1993; a series of 'Notes about the Geology and Palaeontology of the Miocene of Lisboa', Antunes et al., since 1960). Azevêdo (1982a, b) mainly studied the Pliocene deposits of Setúbal Peninsula.

The basal filling of the LTB attributable to the Paleogene comprises alluvial endorheic, poorly sorted deposits that can be differentiated into two units: a lower unit (UBS 7), indicating a predominance of sheet-flood deposits, and an upper unit (UBS 8), reflecting channelled flows.

Mainly alluvial fan deposits accumulated, fed from marginal uplifted terrain (Hesperian Massif). These materials crop out in the current margin of the basin, encircling it almost completely. In Lisboa-Setúbal Peninsula, the deposits constitute the Benfica Formation, in the central sector the Monsanto Formation, in the proximal area the Cabeço do Infante Formation, and in the southern border of the basin the Vale de Guizo Formation. They consist predominantly of coarse deposits, conglomerates, interbedded with sandstones,

Table 1 Results of $^{87}Sr/^{86}Sr$ age determinations for the Neogene of the distal sector of the Lower Tejo Basin

Localities	Samp./bed	Ma	$^{87}Sr/^{86}Sr$	$\pm 2\rho$
Quinta da Piedade (Div. I)	2.5	24(±1)	0.7082887	0.000008
	2.2	21.5(±0.2)	0.7084012	0.000008
	1	24(±1)	0.708288	0.000011
Lisboa-Parque Eduardo VII (Div. I)		22.3(−0.7,+0.4)	0.708365	0.000021
Cotovia (Div. I)		22.2(−1.0,+0.8)	0.708374	0.000041
Carnide 8 (Div. I)		19.7(−0.2,+0.3)	0.7085121	0.000011
Lisboa-Av. Uruguai (Div. II)		21.5(−0.3,+0.5)	0.708397	0.000017
Lisboa-Univ. Católica (Div. II)		20.5(−0.2,+0.3)	0.708463	0.000017
Lisboa (Alto dos Moinhos) (Div. II)	6	19.5(−0.1,+0.2)	0.7085228	0.000009
	4	19.5(−0.1,+0.2)	0.7085232	0.000011
Portinho da Costa (Div. II)		18.3(+0.4−0.2)	0.708612	0.000020
Carcavelos	8	19.5(−0.1,+0.2)	0.7085241	0.000011
	6	20.3(−0.1,+0.2)	0.7084817	0.000009
	1	19.7(−0.2,+0.3)	0.7085121	0.000011
Foz da Fonte	16	18.5(+0.2,−0.3)	0.708601	0.000014
	14	18.8(+0.4,−0.2)	0.708582	0.000017
	13	19.0(+0.3,−0.2)	0.708570	0.000016
	7	19.7(+0.3,−0.2)	0.708510	0.000020
	3	19.5(+0.2,−0.2)	0.708539	0.000017
Penedo South	13	17.6(+0.4,−1.4)	0.708658	0.000016
	11	18.0(+0.3,−0.5)	0.708629	0.000017
	10	18.5(+0.3,−0.4)	0.708603	0.000018
	5	19.6(+0.4,−0.2)	0.708513	0.000018
	1	20.0(± 0.4)	0.708509	0.000020
Penedo North	12	14.5(+1.0,−0.7)	0.708801	0.000016
	10	15.0(+1.0,−0.5)	0.708779	0.000016
	9	12.5(+1.0,−2.0)	0.708855	0.000017
	8	17.3(+0.6,−0.5)	0.708672	0.000016
	5	17.8(+0.7,−0.5)	0.708647	0.000016
	1	17.7(+0.7,−0.5)	0.708650	0.000016
Lisboa (Av. Padre Cruz) (Div. IVa)		17.8(± 0.2)	0.7086459	0.000008
Quinta das Pedreiras (Div. IVb)	9	18.1(−0.3,+0.2)	0.7086335	0.000011
	6	18.2(−0.2,+0.3)	0.7086258	0.000014
	1	17.5 (−0.3,+0.2)	0.7086723	0.000010
	0	17.5 (−0.3,+0.2)	0.708672	0.000011
Casal Vistoso (Div. IVb)		17.3(−0.3,+0.2)	0.7086764	0.000008
Lisboa (Chelas) (Div. Vb)		14.7(−0.5,+1.5)	0.708787	0.000018
Lisboa (Ralis) (Div. VIc)		11.6(−0.6,+1.4)	0.708869	0.000011
Ribeira da Lage	16	11.8(+1.6,−3.3)	0.708862	0.000021
	14	11.3(+1.7,−2.8)	0.708878	0.000018
	8	12.7(+0.4,−1.4)	0.708846	0.000016
	7	11.7(+1.3,−1.3)	0.708869	0.000016
	3	12.2(+1.0,−1.2)	0.708859	0.000016
Fonte da Telha (Div. VIIa-b)		5.2(−1.2,+3.1)	0.708980	0.000023

(continued)

Table 1 (continued)

Localities	Samp./bed	Ma	$^{87}Sr/^{86}Sr$	$\pm 2\rho$
Foz do Rêgo (Div. VIIa-b)	Arca	8.7(−3.5,+1.8)	0.708918	0.000017
	Chlamys	8.3(−3.3,+1.9)	0.708925	0.000016
Belverde borehole	348 m	20.6(±0.3)	0.708474	0.000014
	318 m	20.1(±0.3)	0.708501	0.000014
	299 m	19.6(±0.3)	0.708538	0.000017
	149 m	16.2(±0.53)	0.708735	0.000016
	61 m	11.6(±0.75)	0.708862	0.000013
	36 m	11.6(±0.75)	0.708866	0.000016
	21 m	12.3(±0.5)	0.708852	0.000016
	6 m	10.8(±1)	0.708889	0.000014

feldspathic sands and lutites, calcareous crusts and, more occasionally, lacustrine or marsh limestones.

The Atlantic water invaded the basin in the Early Miocene. Since then, the sedimentation in the Lisboa and Setúbal Peninsula regions has occurred at the continent-ocean interface, with oscillations of the shoreline dependent on eustatic variations and tectonic effects. On the peninsula, the inner part of the continental succession clearly reflects the influence of tectonic and climatic controls on the basin's evolution. The spatial distribution of facies—alluvial deposits units intercalated with marine siliciclastic coastlines—and the abundance of macro and micro fossils, provide evidence of the importance of the stratigraphic record. Both the coastal marine and continental deposits have provided biostratigraphical data of excellent quality. The analysis and integration of numerous lithic and biostratigraphical data, such as the first and last occurrences of some taxa of foraminifera, ostracoda and mammals, together with isotopic ages, have allowed a high-resolution chronostratigraphic framework to be established and ten depositional sequences to be characterized, which correspond in part to the third-order cycles of Haq et al. (1987). Further details are contained in the studies of Antunes et al. (1973, 1987a, 1996a, b, 1999b, 2000), Legoinha (2001, 2008) and Pais (2004).

A detrital unit, sometimes very rich in feldspars (arkoses), interspersed with clasts of quartz and quartzite and with lutite layers (Alcoentre Formation) appears in the central sector. This sandy unit change laterally to a clayey unit with some intercalations of sands (Tomar clays) where, in addition, calcareous crusts and lacustrine and palustrine limestones (Almoster limestones) develop.

The Miocene series is terminated by a disconformity, sometimes strongly erosive. Until about the Tortonian, the basin drained from NE to SW by a sandy fluvial system, developed on a braided alluvial plain. In the late Tortonian and Zanclean, the regime became endorheic with sedimentation in the form of alluvial fans from slopes generated by much younger tectonic reactivation of tardi-Variscan faults. In the Piacenzian, with the onset of a wetter and warm climate, the system was again exorheic and the ancestor of the Tejo River became a major anastomosed, high-energy river, particularly after the capture of the drainage network of the Alto Tejo Madrid centered. This was followed by fluvial incision, involving the excavation of deep valleys and the development of a system of river terraces now evident in the morphology of the Tejo valley.

5.2 Distal Sector of the Lower Tejo Basin (Lisboa and Setúbal Peninsula)

The Neogene of the distal sector of the Lower Tejo Basin (LTB) (Figs. 1 and 2) is particularly well exposed in the Setúbal Peninsula. The Miocene constitutes a syncline (Albufeira Syncline) between the northern region of Lisboa and the Arrábida chain. The succession from the Aquitanian to the Tortonian is bounded above and below by a regional disconformity and an angular unconformity, respectively. This reflects a NW–SE compressional context, related to the Betic collision. The Miocene deposits overlie the Lower Cretaceous, the Volcanic Complex of Lisboa-Mafra or the Paleogene (Benfica Fm.), and are eroded by the Pliocene which passes upwards to the Quaternary. Along the southern flank of the Arrábida chain, the Miocene is overthrust by the Lower Jurassic, creating an angular unconformity between the Lower Miocene and the Middle Miocene.

In Lisboa, the thickness of Miocene deposits does not exceed 300 m. In the region of Barreiro and Montijo, the Neogene attains a thickness of about 1200 m, of which 300 m correspond to the Pliocene. To the west, at Belverde (Seixal), data obtained from deep drilling (619 m) indicate only about 590 m of Miocene and 130 m of Pliocene (Pais et al. 2002, 2003).

Several geological studies have led to the characterization of the lithostratigraphy of the Setúbal Peninsula, in particular of the Municipality of Almada. Cotter (in Dollfus et al. 1903-04) established the "divisions" of the Lisboa Miocene, still used today as lithostratigraphic units. Choffat (1950) sketched and compared stratigraphic columns in the Lisboa area. Antunes et al. (1973) defined sedimentary cycles (transgressions and regressions) based on the marine and continental characteristics of the units previously defined by Cotter.

Antunes and Pais (1993) refined correlations between the region of Almada (Setúbal Peninsula) and the inland units of the Lower Tejo Basin, in the Ribatejo region, and characterized seven transgressions (T0-T6), alternating with six regressions (R0-R5). Correlations with the High Tejo Basin (or Madrid Basin) were established based on major sedimentary ruptures, and their chronostratigraphic positions were defined by mammal faunas (Antunes et al. 1987a).

Antunes et al. (1999b, 2000) presented a chronostratigraphic framework of the continental and marine units from Lisboa to the Setúbal Peninsula. Ten depositional sequences were defined bounded by erosional contacts related to transgressive surfaces. Several paleogeographical models were drawn.

Both paleoclimatic and paleogeographic changes have been recognized. In marine environments, tropical conditions prevailed at least from the Aquitanian, with the establishment of coral reefs until the early Burdigalian. The warmest conditions (comparable to the Gulf of Guinea current) were reached in the late Burdigalian and Langhian. Subsequently, the water temperature was similar to that of the Moroccan coastal current. Fauna and vegetation assemblages show alternating episodes of humidity and aridity (Antunes and Pais 1983; Pais 1986; Lauriat-Rage et al. 1993).

5.3 Paleogene

The distal sector of the LTB corresponds to the area affected by Neogene marine transgressions. The first of the deposits that filled the LTB is the Benfica Formation (Zbyszewski 1963). It consists of continental deposits, reaching about 400 m in thickness. Lithostratigraphic studies, including Choffat (1950), report an Oligocene age, given the formation's position between the underlying basaltic formation, supposedly Eocene, and the Lower Miocene.

More recent studies have been made by Antunes (1967, 1979a), Azevêdo (1991) and Reis et al. (2001). Taking into account new dating of Eocene units and field observations, the Benfica Formation has now been shown to be rather heterogeneous, and by correlation with the fossiliferous Coja Upper Eocene unit, the unit dates from the Eocene (Middle? and Late) to the Oligocene (Antunes 1967, 1979a; Reis et al. 2001).

By analogy and stratigraphic framework, these sediments can be correlated with the Paleogene represented in the N and NW edge of the depression of the Lower Tejo, including the Vale de Guizo Formation from the LTB southern sector, as well as in the regions of Arganil, Coimbra and Nazaré (Mondego Basin), where there are precise Eocene chronological constraints (Bom Sucesso and Côja Formations; Reis and Cunha 1989) (Fig. 18).

The Benfica Formation disconformably overlies the Lisboa Volcanic Complex (Late Cretaceous) (Kullberg et al. 2011b), from which it derived its constituent materials. In some places, the unit contacts directly with the Cenomanian. Above, it is either bounded by disconformity or transitions gradually to marine sediments of the Lower Miocene (Aquitanian).

There are distinct associations of facies that can be differentiated within the deposits. The facies architecture and its vertical and lateral sequence evolution indicate the influence of diastrophic activity, consistent with subsidence in a NE–SW direction and appearing to have been more intense at the top. The differentiation of two episodes separated by a depositional disconformity is attributable to an orogenic phase, a pattern which is found in corresponding deposits in other regions (Reis and Cunha 1989).

The clay fraction is rich in paligorskite and smectite. The smectite is associated with higher levels of detritus, while paligorskite is common in carbonates and mudstones (Azevêdo 1991).

Antunes (1979a) interpreted the lithologies of the formation and recognized three informal units separated by disconformities: unit C (>120 m thick); unit B (about 206 m thick); and unit A (about 106 m thick). Unit A was correlated with the Eocene, while units B and C were assigned to the Oligocene.

Reis et al. (2001) recognized four facies associations. The first association is composed of siliciclastic conglomerates and sandstones, with concave oblique stratifications, sandy-lutite matrix, and elements of quartz, quartzite, lidite, shales and feldspathic clasts of greenish colour. These clasts originated in the Hesperian Massif or were reworked from Mesozoic detrital deposits. Locally, red or whitish marls occur due to the presence of carbonates. Occasionally, Mesozoic limestone clasts appear in the conglomerates. This association corresponds to the unit A of Antunes (1979a) and to the first and lower part of the second beds of Choffat (1950).

The second association is composed of siliciclastic conglomerates and sandstones with carbonate cement, is brick-red, pink, or clear green in colour, and corresponds locally and laterally to the horizons of white nodular micrite sometimes referred to as Alfornelos limestone. The deposits represent lacustrine and swamp facies and have undergone pedogenetic and diagenetic processes. In the region of Loures, carbonates up to around 10 m thick indicate deposition in small endorheic lakes (Azevêdo 1991).

The third association is represented by sandstones and reddish lutite layers and beds containing rosy carbonate concretions. The association includes interbedded decimetre-thickness lenticular bodies and channelled micro-conglomeratic sandstones with coarse, mainly siliciclastic, elements, and more rarely limestone (from the Cretaceous and the erosion of some carbonate crusts developed). The lenticular channelled bodies do not exceed 10 m in width and 2 m in thickness. Basaltic elements are not present. The association corresponds to the occurrence of episodes of sedimentation of fine materials and to the development of intercalated horizons of reddish calcareous crusts. This facies association is equivalent to the unit B of Antunes (1979a), and includes the upper part of the second and third beds of Choffat (1950). The association disconformably overlies the underlying unit, marked by the occurrence of conglomerates with clasts of chert, quartzite, quartz and Cenomanian limestone, bonded by white clay at the the base passing to red at the top (6 m); the limestone clasts are increasingly abundant towards the top. The third bed (Choffat 1950) corresponds to a set of reddish to pink marls, with some scattered clasts and carbonate concretions, and is about 200 m thick.

The fourth association is well represented between Carnide and Póvoa de Santo Adrião (Reis et al. 2001). It corresponds to the fourth, fifth and sixth beds of Choffat (1950) and to unit C of Antunes (1979a). The deposits are tilted about 10° and reach about 120 m in thickness. Coarse sandstones and conglomerates cemented by a reddish clay matrix occur, and contain fragments sourced from the Jurassic St. Pedro limestones and the Ramalhão Schists, and from Cretaceous basalts, sandstones and greenish marls. The set is topped by substantial conglomerates with clasts of quartzite, limestone and quartzite, in bodies of metric thickness alternating with sandstones and red and brown lutites, with nodular or crusted carbonate horizons, and varying in their degree of maturity. The conglomeratic bodies have a marked tabular geometry. The abundance of basalt clasts is variable but seems to be closely related to the occurrence of limestone clasts. The limestone clasts become progressively more abundant and larger towards the top of the succession, corresponding to variations in local supply processes following the NE–SW rupture that affected the Mesozoic units. Spasmodic episodes of deposition occur in channels with high energy and sprawling geometries, and are interspersed with long periods of inactivity and pedological alteration with the formation of carbonate-crusted horizons. Azevêdo (1991) recognized a drainage orientation to the SE.

The succession is similar to that found in the area of the Setúbal Peninsula. It differs, however, by the presence of the Senhora das Necessidades limestone on top of the succession; the limestone unit reaches about 150 m in thickness and forms a continuous band from the Fonte do Sol to the SW of Vila Nogueira de Azeitão. The unit constitutes limestones (particularly "mudstones", ~86% carbonate) that are generally white, with quartz grains, and sometimes with poorly defined stratification. Paligorskite dominates the fine fraction. Fenestrated structures and rhizoconcretions are common, and there are indications of pedogenesis. Zbyszewski (1964) and Zbyszewski et al. (1965) cite freshwater and terrestrial gastropods [*Lymnaea pachygaster*, *L.* aff. *dilatata*, *L. (Stagnicola) syrtica*, *Planorbis (Coretus) cornu* var. *mantelli*, *P. (C.) cornu* var. *solidus*, *Helix ramondi* group, *Helix* sp.]. Azerêdo and Carvalho (1986) refer to fragments of ostracods and charophyte oogonia. The limestones of the Quinta da Marquesa, with oogonia of *Chara* sp. *Chara* aff. *dollfusi*, *Psilochara* sp. and gastropods may be equivalent to the those of Portela Necessidades (Setúbal Peninsula).

5.4 Neogene

The lithostratigraphic base for the Neogene has been defined by Cotter (in Dollfus et al. 1903-1904; Cotter 1956) for the region of Lisboa and Almada, and to the south of Fonte da Telha. The defined lithostratigraphic units (Divisions, with the corresponding formations) remain valid and in use. However, a new stratigraphic framework has since been constructed based on the characterization of depositional sequences and chronostratigraphic adjustments. Ages obtained using planktonic foraminifera, small mammals and Sr isotopes have provided good chronostratigraphic control and have enabled the positions of the different sedimentary disconformities to be calibrated (Figs. 11, 18 and 19) (Antunes et al. 2000; Legoinha 2001; Pais 2004).

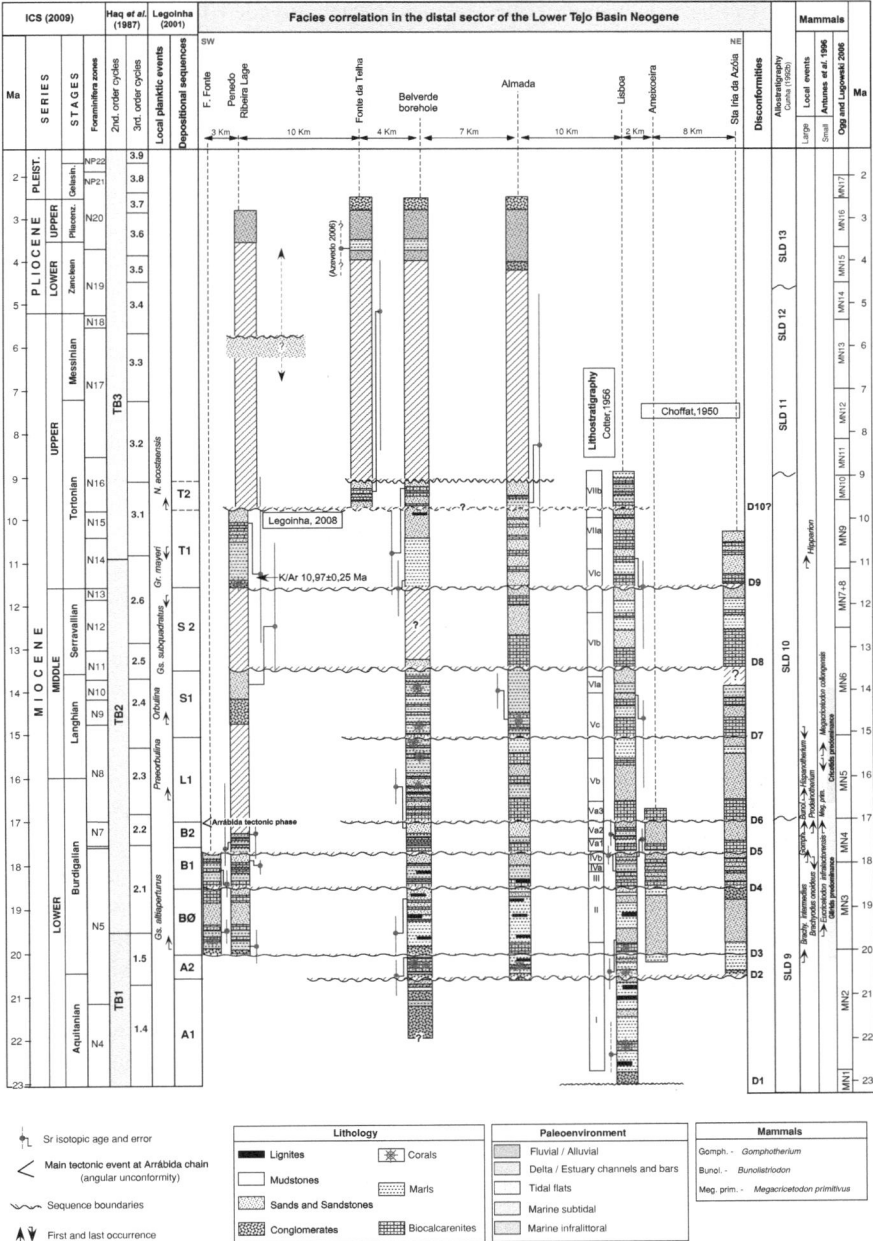

Fig. 11 General stratigraphic framework for the Miocene of the distal sector of the Lower Tejo Basin (modified from Antunes et al. 1998, 2000; Legoinha 2001, 2008; Pais 2004)

In the Aquitanian, water of the Atlantic Ocean invaded the basin. The first entry of the sea was from the south, defining a narrow penetrative gulf. The uplifted terrain of the Arrábida Chain, related to the Burdigalian tectonic inversion (Choffat 1908; Ribeiro et al. 1990a) had not yet formed, but some relief had emerged, inherited from diapiric domes active during the Late Cretaceous-Paleogene (Kullberg 2000; Kullberg et al. 2000). In this morpho-structural framework, a barrier reef was generated that trended N–S and extend at least from the region of Belverde (Seixal) to Lisboa. In the Burdigalian, the sea crossed the western front between Lisboa and Espichel. The Arrábida Mountain formed an island after their uplift in the late Burdigalian (~ 17 Ma). Several gulfs originated from the penetration of the sea into the basin, their extents depending on the eustatic level, amount of subsidence and sediment supply. These gulfs had greater geographic expression (transgression) during the middle Burdigalian (~ 18 Ma), the late Langhian (~ 14 Ma) and the early Tortonian (~ 11 Ma). The major regressive phases correspond to the maximum progradations of sandy units IVb (~ 17.8 Ma) and Vb (~ 16 Ma), reaching the coastline further west than the present. The marine Miocene of the distal part of BBT ends with coastal facies, representing tidal channels and with the occurrence of tempestites (middle Tortonian, ~ 9.5 Ma). It is possible a minor marine transgression may have occured during the Messinian, as indicated by the Rio da Prata sands (Azevêdo 1982a, b).

The evolution of the climate during the Neogene has also been characterized. In marine areas, tropical conditions prevailed. The maximum temperatures were reached in the late Burdigalian and Langhian with values similar to present-day Gulf of Guinea. Thereafter, the temperature decreased to values similar to the present-day Moroccan coast. The faunas and floras indicate alternating wet and drier episodes, with pronounced aridity during the Langhian (Antunes and Pais 1984; Lauriat-Rage et al. 1993; Pais 1981, 1986).

In the late Pliocene there was widespread sedimentary progradation. The ancestral Tejo River transported feldspatic sands (arkoses) from modern-day Spain, which were deposited from the proximal sector and penetrated even into the Alvalade Basin and to the horst of Belverde-Senhor das Chagas, near Alcácer do Sal. In the Setúbal peninsula, further from the source area, the sands are fine, well sorted and practically devoid of pebble beds (Santa Marta Formation). However, at the base of the Santa Marta Formation, gravelly fluvial channels occur, eroding the Miocene marine deposits. In the region of Laranjeiro, these layers include basalt clasts derived from the region of Lisboa, pointing to drainage from the N and NW. To the continental interior (intermediate sector), the Santa Marta Formation is equivalent to the Ulme Formation represented by coarse sands.

In the upper third of the Santa Marta Formation (Fig. 18) a brief transgressive episode is recorded. Brackish waters entered the Setúbal Peninsula area (Azevêdo 1982a, b), as revealed by outcrops of clays with gypsum, macro-vegetation remains (including trunks of *Pinus*) and molluscs. Subsequently, fluvial deposition returned to the peninsula, producing conglomerates with clasts of quartzite and quartz, some faceted by the wind, and supplying probable pre-Acheulean lithic industries atributed to the Plio-Pleistocene boundary (Azevêdo et al. 1979; Azevêdo and Cardoso 1986).

5.4.1 Miocene

For the Miocene, ten depositional sequences have been characterized, initiated by a transgressive surface and shallow marine deposits passing upwards to marls with marine microfossils (transgressive system tract - TST) and finishing with highstand progradational deposits (highstand system tract- HST). These sequences can be related to the third-order eustatic cycles (Antunes et al. 2000; Legoinha 2001; Pais 2004). These depositional sequences include the divisions classically defined in the Lisboa-Almada region by Cotter (1956) (Figs. 11, 18 and 19).

5.4.2 Depositional Sequence A1 (Aquitanian)

Div. I (Part)-Prazeres Clays

The sediments comprising the first sequence have been included in Division I (Cotter, in Dollfus et al. 1903–1904, 1956). They overlie the Paleogene Benfica Formation, the Volcanic Complex of Lisboa or, in some areas, the Cretaceous. The passage of the Oligocene to the Miocene is gradual in some places, but in others is represented by a disconformity or angular unconformity.

The A1 sequence begins with conglomerates that pass upwards to marly deposits and limestones with corals (Antunes and Chevalier 1971; Chevalier and Nascimento 1975) followed by lignite beds and glauconite sands (Choffat 1950; Cotter 1956). This sequence corresponds to the first, second and third parts of Division I as described by Cotter (1956). It seems to be confined to Lisboa.

Ostracods indicate an Aquitanian age: *Aurila (C.) peypouqueti,* and *Hermanites ruggierii* are exclusive to Division I (Nascimento 1988, 1990, 1993). $^{87}Sr/^{86}Sr$ dating (Parque Eduardo VII) yields an age of 22.3 Ma (+0.4,−0.7) (Table 1). Glauconite K/Ar ages provide values between 19 and 24 Ma (Antunes et al. 1973) (Fig. 21).

5.4.3 Depositional Sequence A2 (Aquitanian-Lower Burdigalian)

Div. I (Part)-Prazeres Clays

Clayey and marly environments corresponding to coastal lagoons are predominant. In the Lisboa region, biohermas (corals, bryozoans) were deposited, followed by mudstones. Laterally, there are layers of gypsum and coal with plants (Pais 1981, 1986) which provided the mammal fauna of Horta das Tripas. These layers are overlain by fine to medium micaceous sand, also containing mammals remains (Universidade Católica fauna, Antunes and Mein 1986). Sandy mudstones occur at the top with channels filled with oyster shells. These layers have also yielded mammals fossils (Avenida do Uruguai fauna; Antunes and Mein 1986). The maximum thickness of the whole sequence reaches ∼45 m.

Ten kilometres from highway AE 1, an outcrop of gravelly layers containing oyster shells has provided a small mammal fauna, atributed to MN2b (MN3?) (Figs. 11, 18 and 19) (Antunes and Mein 1992; Antunes 2000), with *Cainotherium* sp., *Lagopsis spiracensis*, *Ligerimys antiquus*, *Pseudodryomys simplicidens*, and *Heteroxerus rubricati* (archaic), among others. It is the oldest site with Neogene mammals known in Portugal.

5.4.4 Depositional Sequence B0 (Burdigalian)

Div. II-Estefânia Avenue Sands With *Chlamys pseudo-pandorae*

This sequence varies in thickness from 35 m in Lisboa to 2–3 m in Almada, where it is represented by biocalcarenites and sands of the upper part of the unit cropping out at the base of the escarpment of the Tejo River. In Portinho da Costa, there are layers rich in oyster shells. Sands with mud lenticles contain plant remains at the base of the Miradouro of Almada section, slightly above the level of the Tejo River (the deposit is covered by landfill). *Hylodesmum podocarpum* predominates, together with *Carpinus* cf. *orientalis*, *Salix* sp., *Populus serrulatus* and *Ulmus* sp. These suggest a mixed mesophytic forest growing on flat terrain and developing under a subtropical climate (Pais 1981).

The set of the Estefânia Sands belongs to a single depositional sequence. The basal disconformity (D3), which is not exposed in the Almada Municipality, has been dated at 20 Ma (Antunes et al. 2000; updated, ICS 2009) (Figs. 11, 18 and 19).

5.4.5 Depositional Sequence B1 (Burdigalian)

Div. III-*Banco Real*

This sequence is represented by biocalcarenites with an abundant detrital fraction and casts of molluscs. In Portinho da Costa laminated sandstones with fragments of molluscs occur, passing upwards to very fine sands and clayey silts, grey in colour, of unit IVa. The deposits crop out in a narrow band along the cliffs of the Tejo River. The estimated thickness varies between 5 m in Lisboa and 10 m in the region of Almada.

The "Banco Real" (Unit III) overlies the deposits of sequence B0 by a disconformity of regional importance (D4), corresponding to a transgressive surface dated at 18.6 Ma (Antunes et al. 2000, updated, ICS 2009) (Figs. 11, 18 and 19).

Div. IVa-Forno do Tijolo Blue Clays

The deposits in this division contain a set of fine clayey sands, with blue-grey pyrite (euxinic facies), and with molluscs (including *Pereiraia gervaisi*), fish and abundant microfossils (calcareous nanoplankton, dinoflagellates, foraminifera, and ostracods). The division corresponds to the most important Burdigalian transgression. Calcareous nanoplanktons from the Palença (Almada) section (Fig. 12) indicate zone NN4 (Fonseca 1977).

The associations of ostracoda from the Cristo Rei section (Fig. 14) (Nascimento 1988) indicate an infralittoral environment. The fauna are typical of warmer waters, although they may indicate temperatures lower than those of the Aquitanian. The species present in this section are common in the Burdigalian and Aquitanian, in association with thermophilic forms, *Cnestocythere* and *Pokornyella* (Nascimento 1988, 1990). This is consistent with the occurrence of hermatypic corals in the Lower Miocene of the Lower Tejo basin (Antunes and Chevalier 1971, Chevalier and Nascimento 1975).

The top of Unit IVa shows a decrease both in salinity (plant macrofossils, *Cerithium* and other molluscs) and in depth. The deposits are well exposed in the cliffs of the Tejo River, a little west of Cacilhas to Trafaria. They measure 25 m in thickness and tilt about 7° to the S in the region of Almada, while in the Lisboa/Loures region the thickness does not exceed 11 m.

Div. IVb-Quinta do Bacalhau Sands

These are relatively thick (about 35–40 m) prograding deposits of fluvial arkosic sands, with beds of mudstones corresponding to channels and muddy deposits of floodplain and delta environments. The predominant color is yellow. In the fluvial and deltaic arkosic deposits, mammal associations have been collected at Quinta do Narigão, Quinta da Noiva, Quinta da Carrapata and Pote de Água (lower part) and Cristo Rei (top), assigned to the MN4 zone, which includes *Gomphotherium* and the last *Brachyodus onoideus* (Antunes 1984, 2000; Antunes et al. 1996b). The unit is well exposed in Quinta da Barbuda (Fig. 13).

Oysters are common in the deposits. At Lumiar, the uppermost levels are dark grey clays, rich in gypsum, with oyster banks intercalated. The Quinta do Bacalhau sands are well represented along the cliffs between the Tejo River and Cacilhas-Trafaria.

The sucession represents the regressive deposits and low-stand of the depositional sequence B1. In the section of Cristo Rei (Antunes et al. 1999c) sand layers (Fig. 14), remains of animals and clay lenticles with numerous impressions of plant leaves are well exposed. These deposits correspond to this episode of regression (between 17.6 and 17 Ma) and to a distal position (when compared to the typical fluvial facies represented in Lisboa) closer to the axis of the basin.

Fig. 12 Almada belvedere
section (modified from Pais
1981)

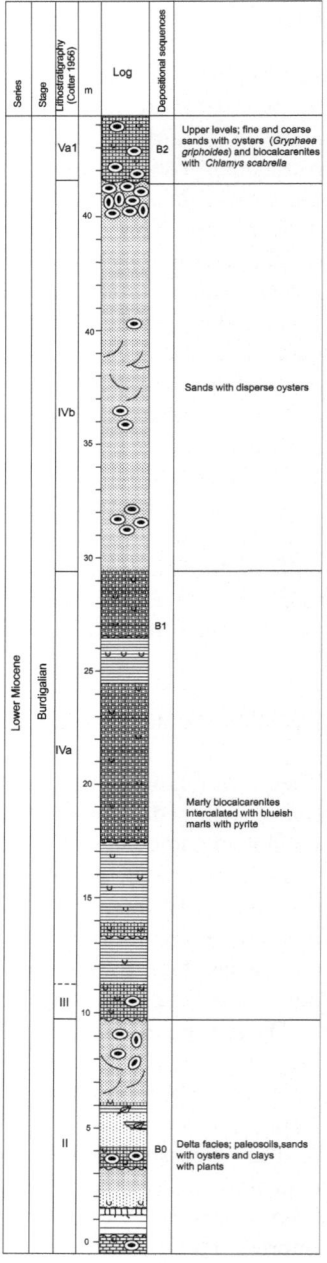

Fig. 13 Quinta da Barbuda
section (modified from Pais
1981)

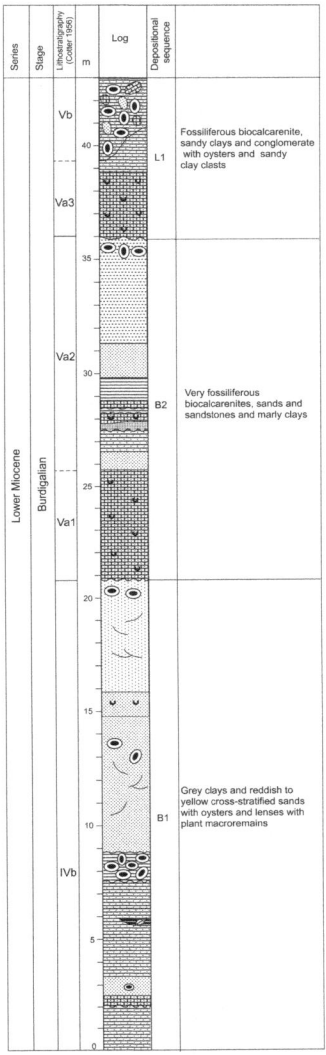

Geological and paleontological (plants, fish, crocodiles and terrestrial mammals) analyses of the section have allowed the reconstruction of various environments: estuarine channels, the upstream areas occupied by brackish waters with banks of oysters (*Gryphaea gryphoides*) passing to the fresh waters, flanked by wetlands and subtropical low mountain forests, which developed under a warm-temperate and rainy climate. In seasonally dry environments there was little dense forest, shrub cover or steppe.

The plant site located in the middle of the section, under the monument to Cristo Rei (Fig. 14), has provided various species, including (Antunes et al.

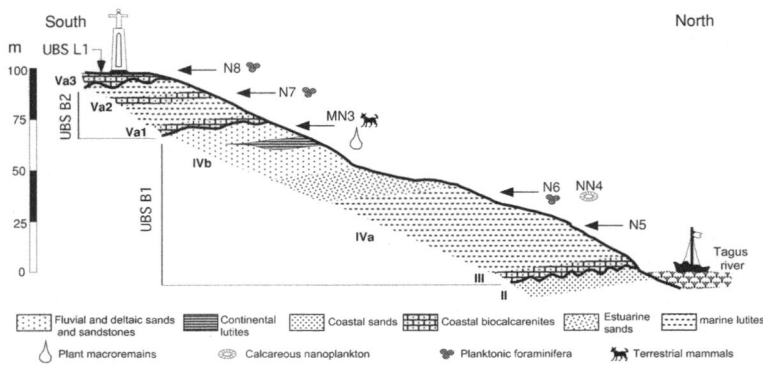

Fig. 14 Cristo Rei Miocene section (modified from Antunes et al. 1999c)

1999c): *Lygodium gaudinii, Comptonia acutiloba, Myrica* cf. *lignitum, Populus serrulatus, Ulmus bronnii, Zelkova zelkovaefolia, Sapindus falcifolius, Magnolia oedipa, Daphnogene polymorpha,* Cf. *Engelhardtia orsbergensis, Hylodesmum podocarpum* and *Gleditsia knorrii.* The state of preservation is reasonable, but unfortunately the cuticles were not preserved. *Hylodesmum podocarpum* is particularly abundant, followed closely by *Populus serrulatus.* The remaining species are reduced to two or three exemplars, and some have only one.

The set includes forms with clearly Asian affinities (*Populus serrulatus, Zelkova zelkovaefolia, Engelhardtia orsbergensis,* and *Sapindus falcifolius*) and, on the other hand American affinities (*Magnolia oedipa* and *Comptonia acutiloba*). In addition, *Gleditsia knorrii* and *Myrica lignitum* seem related to African forms. These species suggest a temperate warm to sub-tropical climate. The abundance of leaves with non-entire margins suggests the existence of well-defined seasons and seasonal high humidity.

From the standpoint of leaf physionomy, the IVb division set of macro-remains collected in the Lisboa region comprises 5.6% of leptophylls, 22% of nanophylls, 61.1% of microphylls and 11.1% of notophylls. These values correspond to low altitude mountain forests developed under warm-temperate climate with high humidity. Regarding the cutting of the leaves, the majority is entire (55%), which indicates, by comparison with the existing vegetation, that the vegetation grew in subtropical forests of low mountains, either in a warm-temperate and rainy climate or in a climate characterized by seasonal dryness.

From its stratigraphic position, just below the limestone level of unit Va1, the paleontological site of Cristo Rei can be dated from the middle to late Burdigalian. Small mammals in two layers immediately above the bed with plants have allowed it to be attributed to the MN4 zone (late Burdigalian, about 17.8 Ma) (Antunes et al. 1999c). It seems a little younger than those of Quinta do Bacalhau and Campo Grande (Quinta do Fidié), the lowest in Div IVb and approximately contemporaneous with those from Quinta das Pedreiras and Quinta da Barbuda, in the Lisboa region. These deposits include *Gomphotherium* and the last *Brachyodus onoideus* (Antunes 1984, 1990a; Antunes et al. 1996b). Fish are common, mainly selaceans (*Pristiophorus* sp., *Megascyliorhinus miocaenicus, Scyliorhinus joleaudi, Paragaleus pulchellus, Hemipristis serra, Galeocerdo aduncus, Rhizoprionodon* sp., *Carcharhinus priscus, Sphyrna* sp., *Rhynchobatus pristinus, Rhinobatos* sp., *Dasyatis* gr. *centroura, D.* gr. *gigas, Myliobatis* sp., *Pteromylaeus* sp. and *Rhinoptera* sp.). Crocodiles are represented by *Tomistoma lusitanica* and *Gavialis* sp. The following mammals have also been identified: Artiodactyla - *Cainotherium miocaenicum, ? Lagomeryx ruetimeyeri;* Lagomorpha - *Lagopsis peñai*; Rodentia - *Pseudodryomys simplicidens, P. robustus, Peridyromys murinus* or *biradiculus, Ligerimys antiquus* (evolutionary stage similar to Beaulieu and Moratilla localities), and *Heteroxerus* sp. (Figs. 11, 18, 19 and 22).

5.4.6 Depositional Sequence B2 (Upper Burdigalian)

Div. Va1-Casal Vistoso Limestones with *Chlamys scabrella*

This unit comprises a carbonated sandy bed, sometimes coarse, very rich in molluscs and rodophyte seaweed, with thickness between 3 and 12 m. The base of the unit rests on a surface of disconformity (D5) corresponding to the transgressive surface that marks the beginning of depositional sequence B2 at 17.7 Ma (Antunes et al. 2000; updated, ICS 2009) (Figs. 11, 18 and 19).

Div. Va2-*Placuna miocaenica* Sands

This sandy unit corresponds to the regressive phase of the depositional sequence. Yellow fluvial sands with pebbles, and sandy clays with plants and oysters, represent the Div. Va2. Pyrolusite impregnation is common, giving the deposits a black colour. Upwards there are sands, partly eolic, associated with thin beds of clay, which could correspond to coastal dunes and deltaic environments. The unit attains a thickness of about 25 m.

The fluvial sands with pyrolusite in the Lisboa region have provided mammals in several localities: Quinta das Pedreiras, large and small forms; Quinta do Pombeiro, important for its small fauna; and Quinta da Conceição (Alto de São João), more marine facies with little material.

The layers contain the same fauna characterized by the arrival of *Bunolistriodon, Dorcatherium, Gaindatherium, Prodeinotherium, Megacricetodon primitivus* and *Democricetodon hispanicus*, allowing the unit to be assigned to the MN4 zone (Antunes 1984, 1990b, 2000; Antunes et al. 1996b).

The sands have provided many other vertebrates: reptiles (crocodiles, Squamata—including snakes and lizards—and chelonians), freshwater or brackish fish (Lates, catfish), and marine fish, sometimes redeposited. This is a strictly thermophilic association, which is consistent with the presence of *Placuna miocenica*.

In the Almada Municipality, the unit is represented by fine clayey sands and marly siltstones revealing a greater marine influence than in Lisboa, and sometimes containing rich micro-fauna deposits.

Concerning ostracods, the lower layers present *Cytherella* (*Cytherelloidea*) *jonesiana* and *Cytheretta* (*Cytheretta*) *rhenana rhenana*. The upper beds include *Aurila* sp., *Miocyprideis fortisensis*, *Cyamocytheridea strigulosa* and *Olimfalunia* gr. *plicatula*. The last appearance is recorded of *M. fortisensis* in the Lower Tejo Basin, where it can reach 80% of the population. The assemblage is suggestive of oligohaline environments and is a local indicator of the end of the Burdigalian and of warm water (Nascimento 1988, 1990, 1993).

The unit is represented in the upper part of the cliffs on the left bank of the Tejo River, extending to Trafaria. It is well exposed in the upper part of the Cristo Rei section and in the area of the Instituto de Estradas de Portugal.

In the Pica Galo (Trafaria) section, a normal magnetic polarity zone has been recognized (Sen et al. 1992). Given the biostratigraphic data and isotopic ages, the sequence correlates with the zone C5Cn. The $^{87}Sr/^{86}Sr$ age of the base of the sequence is 17.7 (+0.7,−0.5) Ma and the top is 17.3 (+0.6,−0.5) Ma (Figs. 11, 18 and 19).

5.4.7 Depositional Sequence L1 (Upper Burdigalian, Langhian and Lower Serravallian)

Div. Va3-Musgueira Limestones with *Chlamys scabriuscula*

This sequence disconformably overlies unit Va2 on a transgressive surface. It is represented by biocalcarenites, which are white, sometimes yellowish in colour, sandy (often coarse), and very rich in molluscs and rhodophyte algae (although less abundant than in division Va1). Upwards, the biocalcarenites pass to sandy marls with abundant fragments of *Schizaster scillae*. On top, locally there are arkoses [about 5 m thick at Quinta das Rosas (Monte de Caparica) section (Antunes et al. 1996b)].

The Musgueira limestone reaches about 5 m thick and crops out in the upper escarpment bordering the Tejo River, extending to near Trafaria.

The basal discontinuity (D6) with an age of 17.0 Ma corresponds to an important tectonic phase in Arrábida (Antunes et al. 2000; updated ICS 2009) (Figs. 11, 18 and 19).

5.4.8 Divs. Vb e Vc (part)-Vale de Chelas Sands (Vb) and Quinta das Conchas Limestones with Spathic Fossils and *Anomia choffati* (Part Vc)

These sequences are represented by yellowish thin feldspar sands, incoherent or weakly cemented, sometimes coarse and compact, with cross-bedding, constituting the lower sands of the Quinta da Silvéria, corresponding to a maximum regressive episode (Antunes and Torquato 1969–70); in the upper part, aeolian sand dunes occur. The deposits correspond to the regressive phase and low sea level of the eustatic depositional sequence L1.

In the Lisboa area there are vertebrate localities in the sands of div. Vb. In the lower part, the deposits of Quinta da Farinheira and Quinta das Flamengas are included within a single designation of Chelas 1, assigned to the MN5 zone. Other contemporaneous deposits are known in Charneca do Lumiar (Quinta da Silvéria, Quinta Grande, Olival da Susana, and Casal das Chitas), and include the first occurrence of the *Hispanotherium* (Antunes 1979b) fauna and the last *Megacricetodon primitivus*.

At the top of the unit is the site Chelas 2 with *Megacricetodon collongensis*, also dated from the MN5 zone (Antunes 1984, 2000; Antunes et al. 1996b). In the Almada region, the deposits crop out in a band at the top of the escarpment of the Tejo River.

The unit includes such foraminifera as *Praeorbulina glomerosa*, *Globigerinoides sicanus* and *Praeorbulina transitoria*, which indicate Blow's zone N8; *Globigerinoides altiapertura* and *Globoquadrina* are rare, whereas *Globigerinoides subquadratus* is abundant. Regarding the benthic forms, *Ammonia*, *Nonion*, *Elphidium*, *Cibicides*, *Lenticulina*, *Textularia* and *Bulimina* are well represented. The association points to deep infralittoral environments, an interpretation confirmed by ostracoda (Antunes et al. 1995).

That deposit passes upwards to shallow marine sands intercalated with some clayey-marly levels. *Praeorbulina glomerosa curva*, *Praeorbulina transitoria*, *Globigerinoides sicanus*, *Globigerinella aequilateralis*, *Globorotalia praescitula* and *Globorotalia peripheroronda* also indicate zone N8. However, the existence of *Praeorbulina glomerosa circularis* suggests proximity to zone N9 (Antunes et al. 1995).

The ostracoda include *Aurila (Uliczpina) oblonga*, *Ruggieria (R.) nuda* and *Loxoconcha (Loxoconcha) ducasseae*, unknown in the Lower Miocene. The association has clear affinities with the Serravallian and suggests warmer waters than previously and environments close to the infralittoral/circalittoral boundary (Nascimento 1988, 1990; Antunes et al. 1995) (Figs. 11, 18, 19 and 23).

5.4.9 Depositional Sequence S1 (Langhian-Lower Serravallian)

Divs. Vc (Part), VIa e VIb (part)-Quinta das Conchas Limestones with Spathic Fossils and *Anomia choffati* (Part, Vc), Xabregas Blue Clays (VIa) and Grilos Sandstones with *Schizaster scillae* (Part, VIb)

This unit contains more silt than clay and owes its name to the predominant shades that it bears, when unaltered, due to the presence of tiny spherules of framboidal pyrite. This feature, incidentally confirmed by the composition of faunal associations, indicates deposition in a euxinic environment, i.e., with deficient circulation of water and oxygen.

A rhythmic succession of beds is apparent. The beds are rich in fossil shells with well preserved fossils: lamellibranchs (especially *Pitar islandicoides*, *Anadara turonica*; *Amussium* pectinids; *Pinna* sp.; and scarce ostreids), gastropods (*Turritella* spp., *Pereiraia gervaisi*, etc.), rare brachiopods (*Lingula*) and vertebrates. Apart from marine mammals (cetaceans), there is a diverse fish fauna lacking strictly tropical species and possessing characteristics of shallow coastal waters. Both associations (selachian and teleosts) include forms of relatively deep environments including fish with light organs (*Myctophum*) (Antunes and Jonet 1970; Balbino 1996).

At this time, the greatest depths recorded in the Lower Tejo Basin were attained (as also revealed by the ostracoda associations), which may correlate with the peak of the Middle Miocene transgression and one of the greatest eustatic variations of the Neogene.

Regarding planktonic foraminifera, the first occurrences of *Orbulina suturalis* (N9) and *Orbulina universa* (N10) are registered. *Globigerinoides subquadratus*, *Globorotalia mayeri* and *Globorotalia menardii* also occur, allowing correlation with Blow's zones N9 to N11.

Concerning the ostracoda, there is the first occurence of *Aurila (Ulicznina) oblonga*, *Ruggieria (R.) nuda*, *R. tetraptera tetraptera*, *Nonucythereis seminulum*, *Pterigocythereis (P.) jonesi* and *Olinfalunia costata* (Nascimento 1988, 1990).

A shell of Pectinidae from the Brielas section has been dated by $^{87}Sr/^{86}Sr$ isotope to yield an age of 14 ± 0.4 Ma. The basal discontinuity (D 7) is placed at 15.1 Ma, and the age of the whole sequence is from 15.1 to 13.5 Ma (Antunes et al. 1973; Legoinha 2001) (Figs. 11, 18, 19 and 24).

5.4.10 Depositional Sequence S2 (Serravallian)

Divs. VIb e VIc (Part)-Grilos Sandstones with *Schizaster Scillae* (VIb) and Marvila Limestones with *Pycnodonta squarrosa* (Part, VIc)

The sediments correlated with the Grilos Sandstones are fine sandstones, barely coherent, light-colored, with abundant fossils but fewer planktonic elements. Pectinidae and Ostreidae are more common, with a lack of ooze-live forms *Pitar islandicoides* and *Anadara turonica*. Bryozoa colonies are abundant and foraminifera and diverse equidae are common, especially *Schizaster* and tiny *Echinocyamus*. There are abundant calcareous tubes of worms *Ditrupa* and *Serpula*.

Collected among the vertebrates are cetaceans, one sea turtle and fish, in a rich association in which pelagic or deep forms are rare. Although expressing a terrigenous influence, albeit less evident than in Lisboa, there is no disputing the occurrence of coastal facies, with depths less than those corresponding to unit VIa.

Regarding the planktonic foraminifera, *Globorotalia mayeri* occurs and *Globigerinoides subquadratus* has its last occurrence, pointing to Blow's zones N13-14 (Legoinha 2001).

For ostracoda, the first occurrence of *Aurila (U.) zbyszewskii*, *Celtia quadridentata*, and *Cytheretta (C.) simplex* have been registered (Nascimento 1988, 1990).

The observed thickness of the sequence is about 10 m. The basal discontinuity (D 8) is dated at 13.5 Ma (Legoinha, 2001) (Figs. 11, 18 and 19).

5.4.11 Depositional Sequences T1 and T2 (Tortonian)

Divs. VIc (Part), VIIa e VIIb-Marvila Limestones with *Pycnodonta Squarrosa* (Part, VIc), Braço de Prata Sands with *Flabellipecten Tenuisulcatus* (Div. VIIa) and Cabo Ruivo Sands with *Chlamys Macrotis* From (VIIb)

The sequence starts with a carbonate layer rich of Pectinidae (and, in particular, ahermatypic coral *Flabellum basteroti*) which, among other places, is exposed in the trenches of the IC 20 highway to Costa de Caparica. Upwards, there are alternations of very fossiliferous fine sands and thin beds of marly and sandy limestones about 20 m thick.

The best outcrops occur in the vicinity of the Fossil Cliffs of Costa da Caparica, towards the south of this locality, particularly the section of Foz do Rêgo, which has been referred to since Dollfus et al. (1903–04). The lithological characteristics allow two units to be differentiated in that area:

(a) Mudstones with *Megacardita jouaneti* and turritelids, and alternations of biocalcarenites and marls of Foz do Rêgo and Fonte da Telha (DS T2); and

(b) Caparica calcarenites, Capuchos sands, and sands with *Chlamys macrotis* (DS T1).

In the lower part of the sands containing *Chlamys macrotis* and mudstones with *Megacardita jouaneti* and turritelids, at Foz do Rêgo there is a layer of massive accumulation of *Chlamys macrotis*, with lamination defined by the loose valves. Gonzalez-Delgado and Andrés (*in* Antunes et al. 1990a) reported that these shell concentrations are related to tempestites. The genesis of the sequence is considered to lie within the context of the highstand systems tract, with the relative rise of sea level balanced by sediment supplied by the prograding delta of the Tejo River (González-Delgado et al. 1993, 1995).

In Foz do Rêgo, fine to medium sands occur, which are very bioturbated and show cross-lamination and stratification. The sediment is white or ochre, and the effects of hydromorphism and deferralitization of micas and overall sediment are observed. The geometry is lenticular. The sands are well sorted, without split of traction, suspension or saltation loads, suggesting that transport conditions were rather uniform with only small fluctuations in kinetic energy. The occurrence of helical bioturbation (*Gyrolithes*), presumably generated by crustaceans, and other vertical burrows suggests that the sedimentation was very slow but almost continuous.

Five metres up in the layer of *Chlamys macrotis*, there is a 1-m-thick concentrated shell layer, with a great predominance of molluscs; the layer has great lateral continuity but thins to the north. It is a composite, multispecific, mixed shell concentration. The upper and lower boundaries are gradual. The upper part of the layer is highly bioturbated and contains diagenetic casts of molluscs. In the lower part, fossil preservation is good, with the presence of growth stages of the more abondant taxa and colour traces. Although tending to horizontal, there is no preferred orientation of *Turritella subarchimedis* shells, the most common gastropod. The fragments are very abundant, many are rounded, and complete specimens are scarce. Scarce infaunal bivalves are preserved in their life position: *Panopea*, *Pelecyora, Circomphalus* and *Atrina*. The most abundant disarticulated shells have no preferred orientation of their commissural planes.

The layer contains 141 molluscan taxa (74 gastropods and 67 bivalves). In a 6 kg sample, 511 individuals of Gastropoda and 830 valves of Bivalvia have been counted (0.5 mm mesh size). Among the gastropods, *Turritella (E.) subarchimedis* stands out for its abundance and represents 55% of the total. It is followed in abundance by *Neverita josephinia* (13%), although many shells are reduced to the umbonal button. Fractured Nassariidae (*Sphaeronassa mutabilis* being the most important) make up 7% of the association.

With respect to bivalves, the most representative families are Veneridae (9 species), Tellinidae (8), and Lucinidae/Cardiidae (5). The most common bivalves are *Corbula gibba, Spisula subtruncata, Clausinella* cf. *basteroti, Nucula nucleus, Tellina distorta* and *Nuculana fragilis*.

Dollfus et al. (1903–04) registered 30 species of gastropoda and 31 of bivalves from Foz do Rêgo. Bourcart and Zbyszewski (1940) reported 23 species of gastropods and 13 of bivalves.

Bioturbation is scarce in the lower part of the bed. In addition to microfauna, there are remains of barnacles, decapod crustaceans and celeporiformes bryozoa. Bioerosion by annelids, clionid sponges, perforating algae and bryozoa is common. Encrustation corresponds to Bryozoa and barnacles; evidence of predation on the shells of bivalves and gastropods is probably due to naticids, muricids and decapod crustaceans.

In relation to the distribution of the megafauna of molluscs in the Atlantic domain, similar deposits occur in the Upper Tortonian of Cantillana (Seville) and in the Lower Pliocene of Huelva. In Cantillana, 40 species of gastropods and 34 bivalves are represented. Sixteen species of gastropods and 17 of bivalves are common to the Foz do Rêgo association (Jaccard index of similarity of 0.24 and 0.28, respectively; Dice similarity index of 0.39 and 0.49, respectively), which shows some similarity (particularly for bivalves) between the two associations. Concerning the Huelva region, where there are 130 recognized species of gastropods (González Delgado et al. 1993) and 99 of bivalves (Andres 1982), 22 species of gastropods and 32 bivalves are common (Jaccard Indices of 0.16 and 0.45, respectively, and Dice of 0.27 and 0.62, respectively). The similarity is lower than for Cantillana and between the gastropods and bivalves.

Although the fauna present clear signs of post-mortem transport, they probably lived in infralittoral environments with warm water, normal salinity, and an unstable bottom. The region was periodically affected by storms that led to the accumulation of shells; bioturbation likely reflects the calm periods. Antunes et al. (1990a) state that the teleost fauna suggests relatively warm conditions, but not strictly tropical as inferred from the absence of both hot water stenothermic sharks and giant barracuda (known in the Burdigalian-Langhian). Moreover, stenothermal warm molluscs such as Terebrids, Olivids and *Ficus* among the gastropods, and *Circomphalus foliaceolamellosus, Linga columbella, Nuculana pella* and *Paphia vetula* among the bivalves, suggest warm conditions. Values of 0.79‰ $\delta^{18}O$ (vPDB) for *Turritella subarchimedis* suggest warm waters too, similar to the warm waters reported in the Pliocene of Huelva region for the same taxon (González-Delgado et al. 1995).

Cetaceans are well reported (Antunes et al. 1990a, 1992a) and include: isolated teeth of large size kentriodontids, cf. *Macrokentriodon* sp. (Estevens and Antunes 2003), cranial fragments, mandible and vertebrae of baleen whales (Estevens 2005).

The microfaunal characteristics of these sediments have been studied by Sierro (in Antunes et al. 1990a). The planktonic foraminifera are sparse, but representative, and include: *Globorotalia pseudobesa, Globorotalia menardii, Globigerina quinqueloba, Globigerinoides bulloideus, Neogloboquadrina continuosa* (dextral), intermediate forms between *Neogloboquadrina continuosa* and *Neogloboquadrina acostaensis* (all dextral), and *Orbulina universa, Orbulina suturalis, Globigerina bulloides* and *Globigerinoides quadrilobatus*. To the previous list, Legoinha (2001) added *Globigerinoides mitra, Globigerinoides obliquus, Sphaerodinelopsis disjuncta, Globigerinopsis aguasayensis* and a few intermediate forms between *Globigerinoides obliquus* and *Globigerinoides extremus*. The association indicates a younger age than Blow's zone N15, corresponding to the interval between the disappearance of *Globorotalia mayeri* and the appearance of *Neogloboquadrina acostaensis*.

Pollen analysis has provided dinoflagellates, spores and pollen. Among the dinoflagellates, there is a great abundance of *Achomosphaera* spp., *Batiacasphaera sphaerica, Homotryblium* spp. *Lingulodinium machaerophorum* and *Spiniferites/ Achomosphaera*. Also represented are *Achomosphaera andalousiensis, Impagidinium* sp., *Melitasphaeridium choanophorum, Operculodinium* spp. *Spiniferites mirabilis* and *Tectatodinium pellitum*. A little further south, slightly higher layers have yielded *Homotryblium* spp., *Spiniferites/Achomosphaera, Hystrichosphaeropsis obscura* and *Lingulodinium machaerophorum* (Castro 2006). The set indicates neritic environments with relatively warm waters. Trees dominated the vegetation, the most common of which were pines together with other gymnosperms (*Abies, Cathaya*, Taxodiaceae). The angiosperms were dominated by Fagaceae (*Quercus*), *Myrica* and Oleaceae. There were also *Engelhardia, Castanea, Alnus*, Palmae and Sapotaceae. Herbaceous plants included Amaranthaceae-Chenopodiacae, Compositae, Gramineae and *Armeria*. Spores and herbs occurred, and liverworts were very abundant as well as ferns (Polypodiaceae and Pteridaceae). The vegetation was formed by mixed mesophytic forests with a clear marine influence (Amaranthaceae-Chenopodiaceae, *Armeria*). The abundance of Oleaceae suggests a relatively dry climate with strong seasonal contrasts. The wetter thermophilic elements have low abundances. At the base of the upper bed of the section, there seems to have been a slight increase in humidity as indicated by the marked decline in Mediterranean elements and a small increase in the moister representatives.

A shell of *Chlamys macrotis* and another of *Arca* sp. have been dated by $^{87}Sr/^{86}Sr$ and yielded ages of 8.3 (−3.3 +1.9) Ma and 8.7 (−3.5 +1.8) Ma, respectively (Antunes et al. 2000). However, these ages appear to be anomalously young, as only at the margins of error do they match the chronological interval of zone N15 of 9.8–10.4 Ma. It may be that the beds of the southern sector (DS T2), with *Neogloboquadrina acostaensis*, might belong to N16 (Legoinha 2001). In the Belverde borehole, at equivalent levels, *Neogloboquadrina humerosa* occurs.

The basal discontinuity of the depositional sequence T1 (D9) has an age of 11.6 Ma and that of the T2 sequence has an age of 10.8 Ma (Antunes et al. 2000).

The arkosic sands of the Santa Marta Formation, assigned to the Late Pliocene, overlie the deposits of this unit to the south of Capuchos (Figs. 11, 18, 19 and 25).

5.4.12 Península de Setúbal Region

To the south of the monument to Cristo Rei, it is more difficult to recognize the divisions of Cotter (1956). The environments are marine, and the facies changes related to variations in the shoreline and eustasy are not always as clear as in the Lisboa region. However, it remains possible to make correlations with the divisions of Cotter (1956) a little to the south of Fonte da Telha. Between Espichel Cape and Albufeira lagoon, and on the north and south flanks of the Arrábida chain, other units have been defined, and some of the stratigraphic gaps have been characterized (Manuppella 1999) (Figs. 18, 19, 20, 21, 22, 23, 24, 25, 26 and 27).

5.4.13 Aquitanian

Palhavã Marly Limestones

This deposit is defined in Manuppella (1999). It constitutes the bottom of the unit that Zbyszewski et al. (1965) describe as the "early Burdigalian and Aquitanian Complex".

The deposit corresponds to the first marine levels overlying the limestones of Senhora das Necessidades (Paleogene) and have been dated from the Aquitanian to the early Burdigalian on the basis of the presence of *Miogypsina* aff. *borneensis* and *M.* aff *globulina* or *M.* aff *cushmani* (Azerêdo and Carvalho 1986).

The limestones contain *Venus ribeiroi*, corals and other marine fossils. The unit is represented only in the area of the north flank of Arrábida between Palmela and Venda Nova. In conjunction with the Limestone of Senhora das Necessidades, it forms the southern crests of the S. Francisco and Louro hills, attaining 200 m in altitude.

The deposit can be correlated with the layer containing *Venus ribeiroi* of Division I, of the Miocene in the region of Lisboa, as this may correspond in part to the barrier reef (Figs. 15 and 18).

5.4.14 Burdigalian

Azeitão Mudstones and Marls

This unit, exposed in the region of Quinta da Torre (Palmela), includes biocal-carenites and oyster banks intercalated with clayey-marly layers. It overlies the Palhavã Marly Limestones. Zbyszewski et al. (1965) included these deposits in the early Burdigalian and Aquitanian Complex, corresponding to layers attributed to divisions II, III (Burdigalian) and IV (late Burdigalian) (Figs. 15 and 18).

In the vicinity of the southern flank of the Arrábida chain, coarse deposits occur (sandstones and biocalcarenites rich in quartz pebbles), and the following units are recognized (Antunes et al. 1995):

(a) Yellowish biocalcarenites and fine sandstones, poor in fossils, overlying the Paleogene, and oriented N75°W, 40°N. The thickness is ~ 30 m. $^{87}Sr/^{86}Sr$ ages show 18.8 Ma at the base and 17.5 Ma at the top.

(b) Whitish to yellowish biocalcarenites with rhodoliths, and coarse sandstones, sometimes gravelly, with carbonate cement. At the western end of the Chão de Anixa, they overlie by angular unconformity deposit (a), while to the east by paraconformity. Elsewhere, it overlies by angular unconformity the Middle Jurassic. The thickness is ~ 35 m. The $^{87}Sr/^{86}Sr$ age is 16.5 Ma.

(c) Whitish or yellowish siltstones, poor in fossils. The observable thickness is 8–9 m, and the unit has an age of ~ 16 Ma.

Foz da Fonte and Penedo South Calcarenites and Marls

These deposits crop out on the coastal cliffs of Foz da Fonte and Penedo South, in the southern flank of the Lagoa de Albufeira syncline. Predominantly orange and yellow in colour, the facies are proximal at Foz da Fonte and more distal at Penedo South. The youngest beds are exposed in the upper part of the Penedo South section. The calcarenites and marls contact directly over the Early Cretaceous (Albian) through an erosion surface and slight angular unconformity.

The unit can be subdivided into upper and lower parts by a regional discon-formity (D 4), corresponding respectively to the early Burdigalian transgression (depositional sequence B0) and to Divisions III and IVa of the Lisboa region (depositional sequence B1). The lower part, measuring ~ 12 m thick, begins with a conglomerate including clasts of limestone and Cretaceous rocks in a medium-sized sandy matrix, with abundant mollusc remains. Upwards there are alternations of biocalcarenites, corresponding to coastal marine environments, and clay layers indicating greater depth, beds of fine sand and, again, biocalcarenites, truncated by the regional disconformity (D4).

The disconformity represents the onset of intense erosion, with probable emergence prior to a new trangression. The environmental instability is indicated by discontinuities in sedimentation and by concentrations of *Turritella* and oyster valves. Subsequently, sedimentation stabilized and acquired a cyclical character,

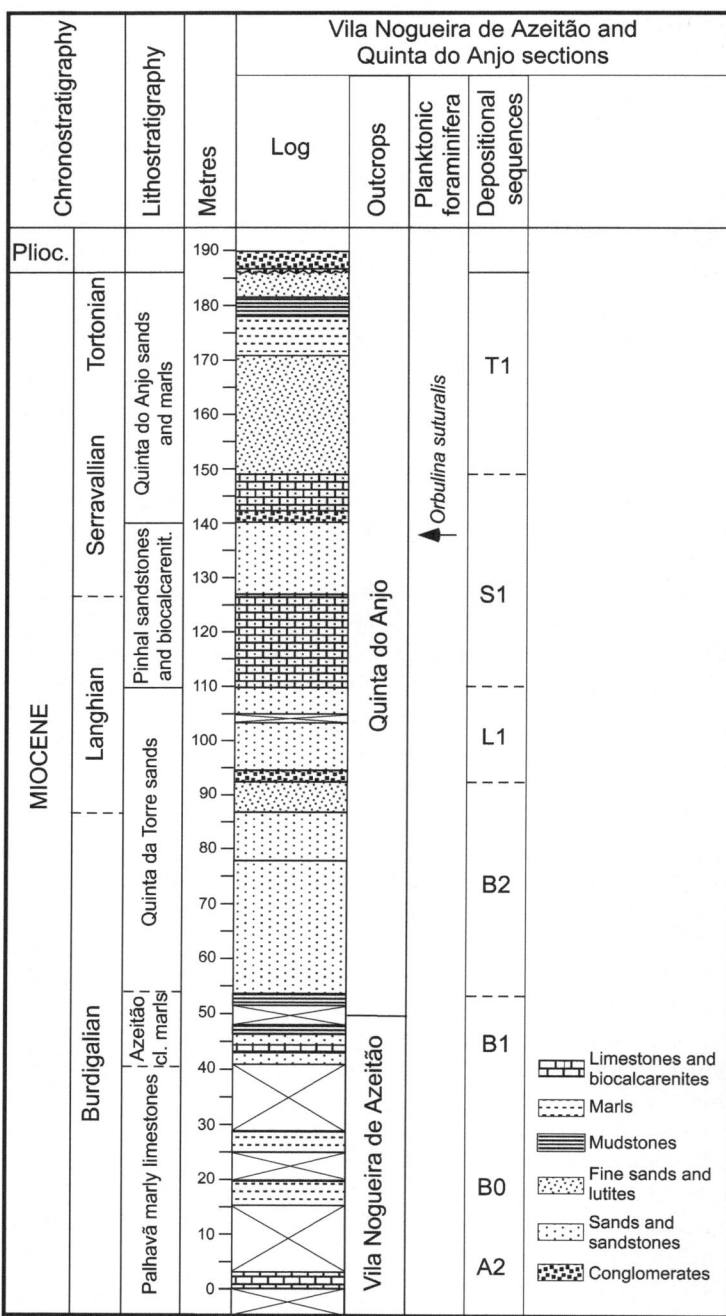

Fig. 15 Stratigraphic log of the northern area of the Arrábida chain at Vila Nogueira de Azeitão and Quinta do Anjo (Legoinha 2001)

expressed by alternations of biocalcarenites and marls (upper part, DS B1). The eustatic maximum in the Burdigalian has been verified as being near the middle of this sub-unit, corresponding to marly beds rich in foraminifera.

Magnetostratigraphic investigations have characterized two normal polarity zones, correlated with magnetic zones C5E and C6, corresponding to Blow's zones N5 to N7 of planktonic foraminifera (Sen et al. 1992). Regarding age control, Antunes et al. (1997c) have indicated $^{87}Sr/^{86}Sr$ ages of ~ 20 Ma for the basal conglomerate (DS B0) and ~ 17.6 Ma for the higher layers of the upper sub-unit (DS B1).

Regarding the foraminiferal biostratigraphy, *Globigerinoides altiapertura* occurs throughout almost the entire unit. The abundance and species diversity of *Globigerinoides* allows the lower sub-unit to be assigned to Blow's N5 zone. In the upper sub-unit, *Praeorbulina* cf. *transitoria* and *Globigerinoides triloba* are more abundant and together with the disappearance of *Catapsydrax unicavus* allows these layers to be designated as belonging to Blow's N7 zone (late Burdigalian) (Legoinha 2001) (Fig. 18).

5.4.15 Langhian-Serravallian

Penedo Glauconitic Deposits

The unit is well exposed at Penedo North, at the northern extreme of Bicas beach. It consists essentially of fine, dark grey, muddy sands. Romariz and Carvalho (1961) worked on the glauconitic levels of the Penedo North section and attributed it to the Late Miocene (Tortonian). Zbyszewski et al. (1965) described the section and correlated it with the top of the late Helvetian. Later, Zbyszewski (1967) considered it as Helvetian divisions VI a–b. More recent studies have clarified the biostratigraphic and chronometric dating of the unit and have characterized the evolution of the isotopic composition of $\partial^{13}C$ and $\partial^{18}O$ (Antunes et al. 1992a, 1995, 1996a, 1997c; Legoinha 2001, 2008), as discussed below.

From the stratigraphic point of view, two particularly significant surfaces of discontinuity can be observed. The first one, at the bottom of the section, is represented as a decimetric conglomerate with rolled molds of bivalves with black phosphate patina, passing upwards to very fine grained sandstones, which are bioturbated and grey in color. The sediments below this discontinuity contain a rich association of planktonic foraminifera pointing to Blow's zone N7 (late Burdigalian). The conglomerate contains *Orbulina suturalis, Praeorbulina* cf. *glomerosa* and *Praeorbulina transitoria,* indicating the middle Langhian (N9). This reveals the existence of a gap in sedimentation correlated with the tectonic phase recorded in Portinho da Arrábida, dated from 17 Ma (Antunes et al. Antunes et al. 1995) and with DS L1 (Fig. 8). The sandstone immediately above contains *O. Universa, Globigerinoides subquadratus* and *Globorotalia* cf. *menardii.* Planktonic foraminifera allow correlation of the bottom and the top of the sandstone to zones N10 and N11, respectively (Legoinha 2001).

Upwards, there is another surface of discontinuity that is strongly bioturbated and overlapped by a conglomerate. In addition to quartz pebbles, the conglomerate contains glauconite and fragments of phosphatized crusts. Vertebrate remains are abundant. Among the selachians, Carcarhinids (*Hemipristis, Galeocerdo* and *Negaprion*-stenothermic forms living in warm waters) predominate. Lamniformes are common, in particular *Isurus*. Sirenids, odontocete cetaceans and one seal are also represented, and a deer astragalus dragged into the sea from the nearby land area has also been found (Antunes et al. 2000).

Overlying the conglomerate, there is a medium-grained, highly glauconitic sandstone. K/Ar dating of the glauconite has yielded an age of 10.97 ± 0.25 Ma (early Tortonian). Among the foraminifera, *Globorotalia* cf. *menardii, G. mayeri, Neogloboquadrina continuousa, Globorotalia scitula* and *Globigerina druryi* have been detected but *G. subquadratus* has not been found. The association indicates Blow's zone N14 (early Tortonian). This highlights a stratigraphic gap corresponding to DS S2, between the regional disconformities D 8 and D 9 (Legoinha 2008) (Figs. 11, 18 and 19).

5.4.16 Langhian-Serravallian

Quinta da Torre Sands

The unit starts with biocalcarenites that pass upwards to fine micaceous white sand, and it is well exposed at Quinta da Torre (Figs. 15 and 18). At Portinho da Arrábida, there are units that are considered correlatives:

(a) Whitish biocalcarenites, fossiliferous, with a thickness of ~76 m and attitude N30°E, 25°SE; the tilt increases to 50°SE;
(b) and coarse sandstones with gravelly beds intercalated with fossiliferous layers. There is local oblique stratification, with structures that allow the inference of currents oriented to the east at the bottom and to the west at the top of the unit. These beds have an estimated thickness in the order of 100 m and a 25°tilt N.

They are overthrusted by the Lower Jurassic. ^{87}Sr/^{86}Sr dating has provided an age of 16.0 Ma (boundary of Burdigalian-Langhian) for both the Quinta da Torre and Portinho da Arrábida units, showing that they are correlative (Antunes et al. 1995; Legoinha 2001).

Pinhal and Castelo de Palmela Arenites and Calcarenites

The unit comprises compact sandstones, whitish in colour, containing pebbles of quartz and quartzite pebbles. At the top there is yellowish fine sand, with scattered oysters. The presence of *Orbulina suturalis* allows correlation with Divisions Vc and VIa of Lisboa despite differences in facies. This unit includes the thick biocalcarenitic beds of the Castelo de Palmela (Legoinha 2001) (Figs. 15 and 18).

5.4.17 Tortonian

Ribeira das Lages Deposits

At Ribeira das Lages, conformably overlying the Penedo glauconitic deposits, there are fine to medium sands, yellow to whitish in colour and containing mica, with many decimetric concretionated layers (Figs. 11 and 18). Towards the top of the section, *Chlamys macrotis* shells became abundant. Zbyszewski et al. (1965) correlated these deposits with the Tortonian.

The association of planktonic foraminifera is characterized by *Globorotalia* cf. *menardii, Neogloboquadrina continuousa, Globigerina apertura, Globigerina druryi, Globigerinopsis aguasayensis, Orbulina suturalis* and *O. universa*. Neither *G. mayeri* nor *Neogloboquadrina acostaensis* are recorded. The association points to Blow's zone N15 (Legoinha 2001, 2008).

The $\partial^{18}O$ curve indicates open marine environments and, in the upper part of the section, a slight increase in water temperature. The decrease in $\partial^{13}C$ in respect of the Penedo glauconitic deposits may be related to the increased influence of continental and/or more oxidizing conditions (Antunes et al. 1996a).

The unit is capped by a strongly erosive disconformity, and is overlain by yellowish, fine marine unfossiliferous sands named the Rio de Prata Sands (Azevêdo 1982a, b). The sands measure only a few metres in thickness and are disconformably overlain by feldspathic sands of the Santa Marta Formation.

Quinta do Anjo Sands and Marls

These deposits start with ochre microconglomerates passing to biocalcarenites and whitish or greyish clayey-marly deposits rich in oysters (Figs. 15 and 18). The microconglomerates define a ridge between Cabanas and Palmela. Zbyszewski et al. (1965, 1967) correlated these deposits with Division VIc and with the Tortonian.

Guarda Mor Conglomerates

To the west of Setubal (Flamenga-Guarda Mor), sequences of clayish deposits and reddish well stratified conglomerates crop out. These are overlain by yellowish fine sand, considered equivalent to part of the sands and marls of Palmela (Azevêdo 1982a, b) (Fig. 18). The clays and conglomerates may correspond to alluvial fans fed from the Upper Jurassic of the Gaiteiros hill.

Choffat (1908) referred to the marls and sands of the Gaiteiros hill, considering it a caspic facies of the Tortonian. The deposit is not represented either in the northern part of the Arrábida chain or in the Lisboa region. Zbyszewski (1964) named it the Flamenga and Lage Red Formation, intercalated with marine units.

5.4.18 Neogene From the Southern Side of the Arrábida Chain

The geology of the Arrábida chain was first studied by Choffat (1906, 1908, 1950), whose investigations highlighted an angular unconformity between what he termed M 1 and M 2. Zbyszewski (1967) subsequently described the corresponding geological sections. Ribeiro et al. (1979) reinterpreted the tectonic scheme of Choffat; the angular unconformity continued to be marked and, in addition, a complex fold (syncline and anticline) was alluded to in the lower unit of the Miocene, which would have connected the mainland to the minor island Pedra da Anixa.

Pais et al. (1991) presented the lithostratigraphy and structural organization of the Neogene deposits of the southern flank of the Arrábida chain. Lithological units were recognized, ordered from bottom to top, as:

(a) At the base, yellowish biocalcarenites and fine sandstones, fossil-poor, overlying the Paleogene. The thickness is ~30 m, with the beds oriented N75°W, 40°N;

(b) whitish to yellowish biocalcarenites with algae concretions, and coarse sandstones, sometimes conglomeratic, with carbonate cement. In the western part of Chão da Anixa, the layers lie on angular unconformity over unit (a), while in the eastern part it overlies the older units by paraconformity. In other places, it overlies the Middle Jurassic by angular unconformity. The thickness is ~35 m;

(c) whitish or yellowish siltstones, poor in fossils. The observable thickness is ~8–9 m;

(d) whitish fossiliferous biocalcarenites with a thickness of ~76 m and attitude N30°E, 25°SE; increasing in tilt to 50°SE;

(e) and Coarse sandstones with conglomeratic beds intercalated with fossiliferous layers; there is local oblique stratification, with sedimentary structures that allow the inference of currents oriented to the E in the lower part and to the W in the upper part. The estimated thickness is ~100 m with a 25° tilt to N. They are overthrusted by the Lower Jurassic.

Structural relations between different units are as follows: Unit (b), which is folded, locally overthrusts units (c) and (d), and unconformably overlies (a); unit (d) contacts unit (c) by an irregular surface; it is not possible to clearly observe the articulation of unit (e) with the other units, but its attitude and lithology are completely different.

Ribeiro (in Ribeiro et al. 1979) reinterpreted the scheme of Choffat (1908) regarding the stratigraphic relations between the Miocene units. Unit (a) at Portinho da Arrábida, locally folded into a closed syncline verging to the SE, is in sequence with the Paleogene, the Cretaceous and the Upper Jurassic, which are in an inverted position, constituting part of the reverse flank of the Formosinho anticline. This is limited to the south, on the continental shelf (Ribeiro et al. 1990), by the main thrust of the Arrábida chain and, to the north, by another thrust that brings into contact the Middle Jurassic and the Miocene. This set forms a duplex structure, probably of kilometric scale. Thereafter, units (a) to (d) were folded into

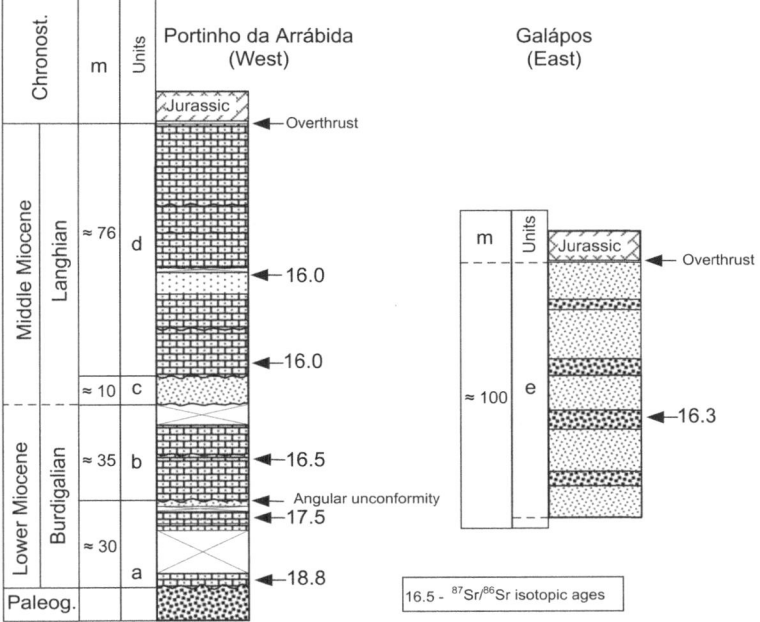

Fig. 16 Stratigraphic columns, lithology and ages of the Miocene for the southern flank of the Arrábida Chain (Antunes et al. 1995)

a sequence of anticline, syncline and anticline. The axes, oriented approximately ENE–WSW, dip to the ENE and pass between Pedra da Anixa and Chão da Anixa, at Praia dos Coelhos, and northwards of this beach, respectively (Fig. 16) (Pais et al. 1991).

Two (and possibly three) tectonic phases are recognized: the first occurring before unit (b), and the second after units (d) and (e). If the overthrust of unit (e) by the Jurassic is not simultaneous with the folding of the sequence, there would be a third tectonic phase corresponding to the reactivation of the thrust that constitutes the northern limit of the duplex referred to (Pais et al. 1991).

Due to the scarcity of planktonic forms, it has proved impossible to obtain dating based on foraminifera. However, it has been possible to obtain isotopic ages of molluscs from the Neogene units of the southern flank of Arrábida: 18.8 Ma (base) and 17.5 Ma (top) for unit (a); 16.5 Ma for unit (b); 16.0 Ma for unit (d); and 16.3 Ma for unit (e) (Fig. 16). The main consequences of these ages are:

- In the southern flank of the Arrábida chain, only part of the Lower Miocene and early Middle Miocene is represented;
- the marine sedimentation started in the middle Burdigalian, probably at the beginning of the Burdigalian transgression, represented in the Lisboa region by Divisions III and IVa;

- the angular unconformity between units (a) and (b) is contemporaneous with the stratigraphic gap recorded in the Burdigalian of the Penedo North section (Legoinha 2008), and marks the occurrence of a tectonic event matching the Iberian Neocastellian or Intra-Aragonian phase;
- unit (e) is a probable lateral equivalent of units (b) and/or (c);
- and it is still not possible to accurately date the second tectonic phase to more recently than 16.3 Ma, which corresponds to the late Serravallian stratigraphic gap of the Penedo North section, and is related to the Iberian Intra-late Aragonian phase.

5.4.19 Pliocene to Pleistocene

The transition Zanclean to Pleistocene represent a great change in the characteristics of the deposited sequences. The first deposits consist of fluviatile conglomerates and coarse sands, almost always arkosic, and often with oblique stratification. These deposits correspond to the *Santa Marta Formation* (Figs. 11, 18 and 19). The first fluvial sedimentation episode related to the ancestral Tejo River is recorded in the lower part of these Pliocene units.

Pliocene sands filled a subsiding depression, corresponding to the routing of the ancestral Tejo River to the ocean. In the vicinity of the Setúbal Peninsula, the sands may reach 320 m in thickness (Pinhal Novo), while on the coast they do not exceed 50 m. Sedimentary structures indicate that the ancestral Tejo was a braided river, with transverse sandy bars (Azevêdo 1997). Clasts of basalt, granite and quartzite of the Lisboa and Sintra regions are common in the pavement of the channels. Azevêdo (1997, 2006) characterized four types of facies (A, B, C and D).

A layer of poorly-preserved molluscs at Fonte da Telha was described by Zbyszewski (1943) and attributed to the Late Pliocene. This layer of a few metres thickness was correlated with fossiliferous beds of Alfeite. The microfauna, however, are unknown. The layer documents a coastal marine association, which does not provide precise dating. In the absence of other information, a Piacenzian age is assumed. The layer is associated with a transgressive episode under a marine influence, with an axial position in the Setúbal Peninsula. There are also layers of grey clays with gypsum crystals, lignite and some diatomites, as well as oysters and casts of *Dreissena*. Locally, there are abundant impressions of leaves and coaly stems.

The Pliocene sands have been exploited by the construction industry. They have been a source of gold, at Adiça, since Roman times (1907) until the end of the 19th Century. Gold grains that reach more than 350 μm in concentrations reaching 3.2 ppm are associated with rare-earth minerals (Salgueiro et al. 2000).

Azevêdo et al. (1979) identified the *Belverde conglomerate* (Figs. 18 and 19) above the sands of the Santa Marta Formation. The conglomerate is whitish, poorly consolidated, and with a sandy matrix. The white clasts, are dominated by quartzite, followed by quartz and, rarely, chert, weathered igneous rocks (possibly

Sintra granite), and sandstones and shales of Ramalhão. At the top, there are abundant ventifact pebbles. The observable thickness is small (2–3 m) and the unit crops out primarily along the coastal cliffs north of the Lagoa de Albufeira. This unit supplied probable pre-Acheulean lithic industries (Azevêdo et al. 1979; Azevêdo and Cardoso 1986). Based on current knowledge, these probable primitive artefacts are compatible with an age between Late Pliocene and Middle Pleistocene, inclusive. The unit may therefore correspond to the Gelasian.

At the top of the sedimentary infill at Setúbal Peninsula, the *Marco Furado Formation* (Figs. 18 and 19) occurs (Azevêdo 1979). It is a gravelly unit with a red, sandy-clay matrix, and reaches 30–40 m thick. The angular clasts are predominantly of white quartz, but quartzite, jasper, chert and shale are also represented. Ferruginous crusts are common, particularly at the top. Sometimes the crusts have been destroyed, leaving debris scattered on the soil surface. In the clay fraction, illite and/or mica and kaolinite are most common (Azevêdo 1982a, b), but there is usually a predominance of illite.

In addition to the units referred to, the Pleistocene is represented by marine terraces that occur at altitudes between 6 m and 90 m. In particular, the marine terraces of Forte da Baralha have provided molluscs. Zbyszewski et al. (1965) refer to the following terraces:

(a) 60 m of altitude. Grey sand immediately N of Baleeira, NNW of Fort da Baralha, with molluscs as described by Dollfus et al. (1903–1904).
(b) 20–25 m altitude. Sands with pebbles.
(c) Between 8–12 m and 6–8 m of altitude. Highest present remnants of ancient beach consisting of consolidated sand with pebbles attributed to Tirrenian II. Both terraces provided fauna.

Elsewhere, there are numerous remnants of marine terrace conglomerates corresponding to 12–15 m and 5–8 m levels attributable to the last interglacial and the beginning of the Würm glaciation (∼ 100,000 years ago, Tirrenian II and III).

The 5–8 m terrace is a narrow platform of marine abrasion with a cover deposit that runs along the open pockets of entry by sea in the coastal cliffs, and the Lapa de Santa Margarida and the Figueira Brava Cave. Zbyszewski (1965) describes the following sequence (from bottom to top) deposited on the 5 m terrace near the Lapa de Santa Margarida:

1. Conglomerates with signficant elements deriving from Paleolithic industries;
2. Red cement conglomerate with fragments of molluscs; and
3. Well-cemented, reddish-brown breccia, with bones of terrestrial vertebrates, Mousterian implements and stalamitic beds.

Zbyszewski and Teixeira (1949) believe that the 5–8 m terrace with molluscan fauna has a similar age to the base of the Devil's Tower cave in Gibraltar (Tirrenian III).

In Lapa de Santa Margarida, stone instruments have been collected from the conglomerate, including a redeposited Abevilian languedocian biface. In the Figueira Brava cave, the industry is Mousterian, with jagged pebbles of quartz and quartzite, and rare pieces of flint. Inside the cave, deposits are not consolidated because they have only weak carbonate cementation; they are protected by a stalagmitic crust.

With the advance of the Würm glaciation, sea level fell. At about 30,000 years ago, the 5–8 m platform and the rock shelters were excavated and remained in an elevated position above the extensive coastal plain. The sea stood about 60 m below the present level (Miskovski 1987). Human communities found good hunting in these areas; the (present-day) caves, positioned above the coastal plain, including the Lapa de Santa Margarida and Figueira Brava cave, constituted natural shelters and contain preserved traces of occupation by humans and animals (remains of Neanderthals, lithic and bone industries, animal bones, and remnants of bonfires). ^{14}C dating and Th/U series ages indicate ~ 30 ka for layer 2 of the archaeological levels explored (Antunes 1990/91).

5.5 Intermediate Sector of the Lower Tejo Basin (Ribatejo and Alto Alentejo)

5.5.1 Paleogene to Plio-Pleistocene

In the northern part of the LTB distal sector (Alto Alentejo and Ribatejo), Neogene sedimentation is related to the migration of the ancestral Tejo River across a vast alluvial plain. Paleogene deposits (*Formation of Monsanto*—on the right bank and *Vale do Guizo Formation*—on the left margin) (Barbosa 1995), which constitute the first infilling of the basin, comprise coarse sandstones and conglomerates arranged in metric fining-up sequences interspersed with lutites, where they develop some calcretes and calcareous rocks (Quinta da Marquesa and Alcanede limestones). In the clay fraction, paligorskite is the dominant component (Figs. 1, 2, 18, 19, 20, 21, 22, 23, 24, 25, 26 and 27).

During the Early and Middle Miocene, sedimentation was mainly fluvial (*Alcoentre Formation* [with several earlier designations; Dias and Pais 2009]). The sandy-clay deposits are sometimes rich in feldspars (arkoses), with pavements of clasts and some lutites interbedded; unconformably overlie Paleogene sediments as well as the Mesozoic and Paleozoic substrate. The existence of some layers with oysters suggests the penetration of brackish water to about 100 km from the current coastline (near Santarém) during eustatic high levels. Some mammals enable correlation with the distal region of the LTB, especially with deposits in the Lisboa area, and particularly for the Middle Miocene and lower Tortonian. The main occurrences are in Vila Nova da Rainha (MN5) (Alberdi et al. 1978); Póvoa Santarém (MN6) (Soulié-Marsh 1978; Pais 1978; Antunes and Mein 1977; Truc 1977; Gaudant 1977); Casais da Formiga (MN7) (Alberdi et al. 1978); Archino (MN9) (Alberdi et al. 1978); and Azambujeira inferior (MN9) (Antunes et al. Antunes et al. 1983). At the top, the *Hipparion* faunas provide ages to match zone

MN9 (lower Vallesian) (Antunes 1979b; Antunes and Ginsburg 1983; Antunes et al. 1992b). In the clay fraction, kaolinite and illite predominate.

The conditions of sedimentation changed in the Late Miocene. On the right bank of the Tejo River, swamp and lacustrine limestones (*Almoster Formation*) pass laterally into clay bodies (*Tomar Formation*) with the development of carbonated crusts, which extend to the existing left bank. They overlie the detrital deposits of the *Alcoentre Formation*. Almoster Formation comprise white to grey limestones; breccia structures are common, and sandy and lutite facies occur. In the clay fraction, smectite and illite predominate. The Tomar Formation consists of reddish-orange mudstones with illite and kaolinite predominance. The ensemble of Almoster and Tomar formations constitutes the Almoster Alloformation (Antunes and Mein 1979; Barbosa 1995).

On the left bank, in the Montargil region, limestones also occur and/or carbonate crusts that developed in the Upper Miocene in the Alcoentre and Tomar formations (Dias et al. 2006, 2009a, b).

Locally, in Rio de Moinhos—Abrantes area, a different sedimentary record of the Almoster/Tomar and Ulme Formations has been recognized (Martins et al. 1998; Barra et al. 2000); it consists of two thickening-upwards sequences (conglomerates to silts). This infilling has been considered the equivalent of the Torre and Monfortinho Formations (Cunha 1996). The units correspond to alluvial deposits (probably from the late Tortonian to the Zanclean) in the SE of the foothills of the Portuguese Central Cordillera.

In the Pliocene, quartz-feldspathic sands (*Ulme Formation*) were deposited (Figs. 18, 19 and 27), extending from the proximal to the distal sector at the Peninsula of Setúbal (Santa Marta Formation). The sands of the Ulme Formation are yellowish to reddish in colour and have medium to coarse texture with high amount of kaolinite and illite in the clay fraction (Azevêdo 1986, 1997; Barbosa and Reis 1989; Barbosa 1995; Barbosa and Reis 1996). Locally there are some fossil plant localities.

At Rio Maior, the sediment accumulation was substantial. The infill (from bottom to top) includes kaolinite fine sand, followed by layers of lignites and diatomites and, at the top, sandy-clay deposits with coarse gravelly layers (Zbyszewski and Almeida 1960). The most complete succession of Rio Maior is situated on the eastern border of the sub-basin, in the area of Espadanal. The sediment thickness reaches 120 m. The diatomites of the Rio Maior basin are rich in microfossils. Andrade (1944) identified different pollens from the Abum outcrop. Diniz (1984) studied two boreholes (F58 and F16) and recognized nine floristic associations designated A to I. These associations represent a complex flora consisting of elements from the Mediterranean, Atlantic and Macaronesian. Based on correlations with synthetic pollen diagrams from northern Europe (Zagwijn 1960) and the Mediterranean (Suc 1984; Diniz 1984, 2001), Diniz concluded that the Rio Maior deposits encompass the whole of the Pliocene, reaching the Quaternary at the top (Suc et al. 1995; Fauquette et al. 1999). However, Cunha et al. (1993) disagree on the basis that a marine level is locally represented at the base of the sucession and it should correspond to the important marine incursion at the Zanclean-Placencian transition.

Vieira (2009) studied the F98 borehole, located in the central area of the Rio Maior sub-basin. This borehole contains a record of almost the entire sedimentary succession, allowing comparison with previous studies. The detailed study of 56 samples provided 22,850 palynomorphs distributed in around 130 taxa. Similar to the study of Diniz (1984), it was possible to recognize, through changes in pollen content, nine generic floristic associations. However, despite the division into these nine floristic associations, taking into account the relationship between pollen content and lithology (floristic changes are also variations in facies), Vieira (2009) recognized three pollen zones (floristic associations I to VII as pollen zone α; association VIII as pollen zone ß; and association IX as pollen zone γ). At the bottom of the core, in pollen zone α (RM 113 to RM 46 samples), there is a predominance of the mega-mesothermal tree species (29%) over the temperate species (15%), and a good representation of spores indicating relative humidity). The mean value of *Pinus* through the zone is 33%. This set corresponds to a mixed forest with plants of subtropical habitat, which presently live in South-East Asia and Florida (USA) and are associated with temperate forests of hardwood trees. In general, the forest association is rich and would have required moist conditions and relatively high average annual temperatures (Fig. 17).

From sample RM43 to RM13 (pollen zone β), there is a decrease in the mega-mesothermal forms (constituting 5% of the total) and an increase in temperate species (to 21%). The percentage of herbaceous plants and spores remain at similar levels, while the percentage of *Pinus* increases to 56%, which may suggest the presence of less dense forest. These data suggest a wet temperature climate. Samples RM7 and RM8 (pollen zone γ) show a lower diversity of palynomorphs. The megathermal and mesothermal forms disappear and the herbaceous plants increase to 26% of the total. Temperate tree species constitute 23% and *Pinus* 49%, and spores fall to 2%. On that basis, a transition to a less forested environment and cooler climate conditions can be inferred.

This new zoning of the pollen content of Rio Maior samples indicates that climatic deterioration took place during the Late Pliocene, and floristic elements were gradually replaced leading to different forest environments (Vieira 2009). Compared to the earlier study of Diniz (1984), Vieira (2009) considers that the Zanclean is not represented and that the deposits are Piacenzian, reaching the Gelasian at the top. Overall, it appears that, in this area, hot and humid climatic conditions prevailed during most of the Piacenzian, in contrast to the climate prevailing in other regions of southern Europe at the time. The warmer, more humid conditions are attributed to the strong influence of the Atlantic Ocean, similar to the effects observed along the present-day Portuguese western coast (Vieira 2009).

In the Late Pliocene, in the proximal sector of the LTB, there was an increase in fluvial energy, leading to the transport and accumulation of gravels (*Almeirim Formation*) extending to the extant littoral coast, where they constitute the *Belverde Formation*. Maximum particle size (MPS) ranges from 45 cm in the

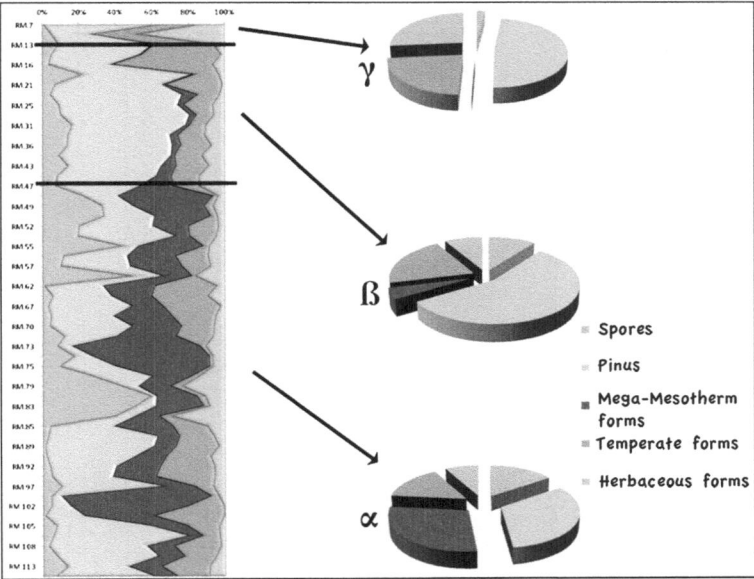

Fig. 17 Subdivision of the simplified pollen spectrum of Rio Maior based on the percentage of dominant groups (Vieira 2009)

most proximal area to 10 cm in the Santarém region. Structures are frequently channelled. In the clay, illite dominates over kaolinite (Barbosa 1995; Barra et al. 2000). The Almeirim Formation conglomerates are overlain by the Vila de Rei conglomerates (Late Pliocene to Pleistocene). The heterometric quartzite clasts in these conglomerates are sub-angular. The matrix is red to orange sandy-pelitic, with the ferruginous cementation suggesting cooler and drier environments. The deposits are arranged in alluvial fans developed near Ordovician quartzite ridges in the north-eastern part of the LTB border (Barra et al. 2000).

5.6 NE Sector (Proximal) of the Lower Tejo Basin

In the north-east sector of the Lower Tejo Basin, the Paleogene crops out extensively but the Neogene is strongly eroded. In the Sarzedas area, the Cenozoic sedimentary infill is preserved and the continuity of outcrops has allowed characterization of the allostratigraphical units that have their sedimentary discontinuities represented at the basin scale (Cunha 1992, 1996, 2000; Cunha and Reis 1992) (Figs. 1, 2, 18, 19, 20, 21, 22, 23, 24, 25, 26 and 27).

Era/Erat.	Peri/Syst.	Epoch/Series	Age/Stage	Faunal units	Ma	Tectonic events	Depositional sequences / Lithost. units (Cotter, 1956)	Lisboa / Almada	Distal sector — Occidental littoral / Setúbal Peninsula (Northern border of Arrábida chain)	Middle sector Ribatejo / Alto Alentejo	Proximal sector Beira Baixa	Unconf. bounded sequences (Cunha, 1992)
Cenozoic	Quaternary	Holocene	Versillian		0,01			Alluvium	Alluvium / Dunes	Alluvium	Alluvium	14
		Pleistocene	Tirrenian / Ionian / Calabrian / Gelasian	Villanian	1,8	Iberomanchega		Terraces	Terraces / Matos Furado Formation	Terraces	Terraces	13
		Pliocene (u.)	Piacenzian		2,6				Belverde Conglomerate / Santa Marta Formation	Almeirim / Congl. de Vila de Rei Fms. / Ulme Formation	Falagueira Formation	12
		Pliocene (L.)	Zanclean	Ruscinian	3,6					Rio de Moinhos congl.	Monfortinho Formation / Torre Formation	11
	Neogene	Miocene (u.)	Messinian / Tortonian	Turolian / Vallesian	5,3 / 7,3	Betic	T2 VIIb Cabo Ruivo Sands / T1 VIIa Braço de Prata Sands / VIc Marvila Limestones		Rio da Prata sands? / Ribeira das Lages deposits / Penedo North glauconitic dep.	Alnostre Fm. / Tomar Fm. / Guarda Mor Sands and Marls Conglomerates?	Silveirinha dos Figos Formation	10
		Miocene (M.)	Serravallian / Langhian	Astaracian / Orleanian	11,6 / 13,7 / 16	"Arrábida" Neocastilian	S2 VIb Grilos Sandstones / S1 Via Xabregas Blue Clays / Vc Quinta das Conchas Limestones / L1 Vb Vale de Chelas Sands / Va3 Musgueira Limestones / Va2 Placuna miocaenica Sands		Penedo North deposits / Fóz da Fonte and Penedo South glauconites calcarenites and marls	Alcoentre Formation		9
		Miocene (L.)	Burdigalian	Agenian	20,4		B2 Va1 Casal Vistoso Limestones / B1 IVb Quinta do Bacalhau Sands / IVa Forno do Tejo Blue clays / B0 III Banco Real / II Estefânia Av. Sands / A2 I Prazeres Clays	Quinta da Torre Sands / Azeitão Mudstones	Pinhal and Castelo de Palmela Sandstones and Biocalcarenites / Quinta do Anjo Sands and Marls		Quinta da Torre Sands	
	Paleogene	Oligocene (u.)	Aquitanian / Chattian		23	Castilian	A1 Facies assoc. IV	Benfica Formation / Facies assoc. III / Facies assoc. II	Palhavã Marly Limestones / Benfica Formation	Monsanto Fm. / Vale de Guizo Fm	Cabeço do Infante Fm.	8
		Oligocene (L.)	Rupelian		34	Pyrenean	Facies assoc. I					7
		Eocene (u./M./L.)	Priabonian / Bartonian / Lutetian / Ypresian		56	Pre-Pyrenean						
		Paleocene (u./M.)	Thanetian / Selandian / Danian		66	Neolaramide / Paleolaramide						

Fig. 18 Correlations between units of the Cenozoic Lower Tejo Basin (updated, Cunha et al. 2009)

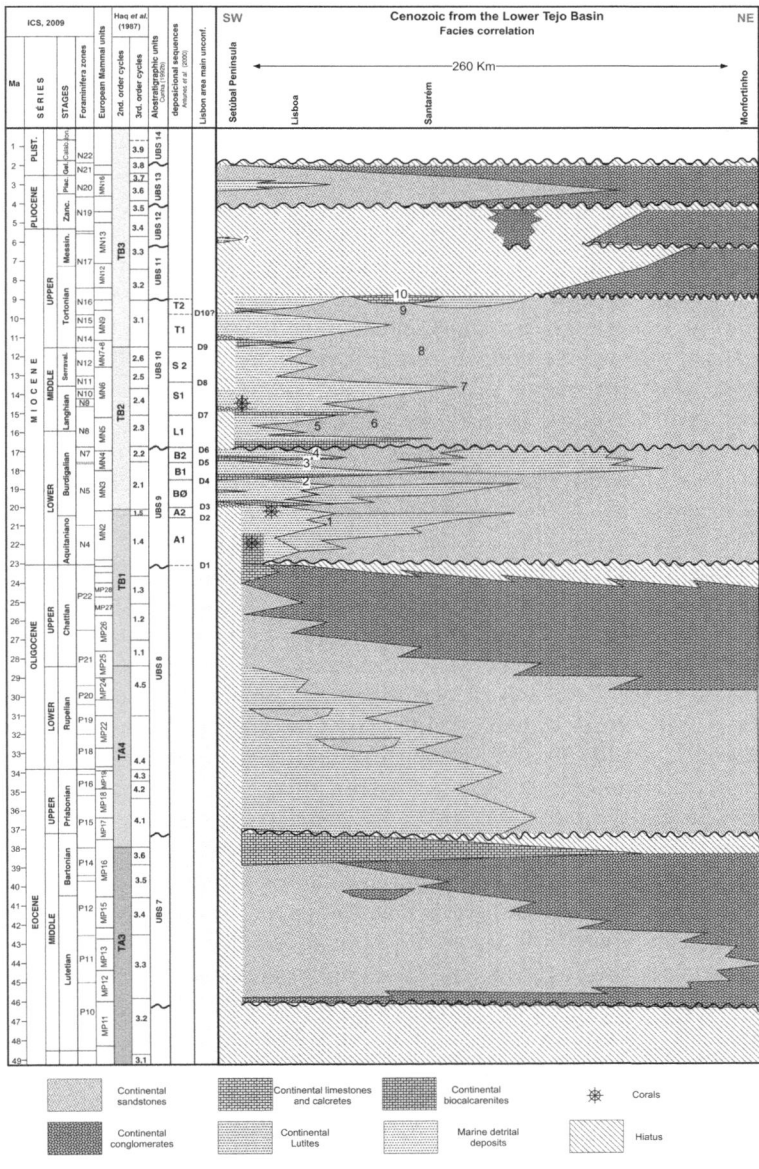

Fig. 19 Facies correlation, according to a profile from the SW (distal sector) to the NE (proximal sector) of the Lower Tejo Basin. *1–10* main mammal localities. Fossiliferous localities: *1* km 10 from the A1 motorway, Horta das Tripas; *2* Avenida Uruguai, Univ. Católica; *3* Quinta do Narigão, Qt da Noiva, Cristo Rei; *4* Quinta do Pombeiro, Quinta das Pedreiras; *5* Chelas 1 (Quinta da Farinheira, Qt. Flamenga, Charneca do Lumiar), Chelas 2; *6* Vila Nova da Rainha inferior; *7* Póvoa de Santarém; *8* Casais da Formiga; *9* Archino, Azambujeira superior, Aveiras; *10* Asseiceira, Freiria de Rio Maior (updated after Cunha et al. 2009)

5.6.1 Paleogene

The sedimentary infill of the Lower Tejo Basin starts with two allostratigraphic units: UBS7 (probably Middle Eocene to Upper Eocene) and UBS8 (Uppermost Eocene to Oligocene) (Cunha, 1992). The lithostratigraphy formally considers this basal sandy-conglomeratic unit as belonging to the Beira Baixa Group and is defined as the *Cabeço do Infante Formation* (Cunha 1987a, b, 1992, 1996) (Figs. 18 and 19). This formation has the same lithological characteristics as in the intermediate sector of the basin, but there the unit was referred to as the Monsanto Formation (Barbosa 1995). Although it has different facies, it is considered to be the same age as the Benfica Formation (distal sector of the basin).

The Cabeço do Infante Formation is rich in hyaline quartz and feldspar grains, with large clasts of quartzite, white quartz and metagreywacke/phyllite. The sediments, poorly sorted have a lutite-smectite matrix. Coarse sandstones and conglomerates predominate, displaying depositional tractive structures (e.g., through cross bedding). They have a greyish or whitish green colour, with violet/red tones in some layers. The top of the formation is usually capped by a silcrete of variable thickness. At the base of the formation, layers cemented by dolomite and rich in paligorskite can reach 45 m in thickness (Cunha and Reis 1989). These accumulations resulted from the replacement of detrital silicates by dolomitic carbonates. Another feature is the occurrence of oxy-hydroxides of iron and manganese in crusts or in dendritic concentrations.

The base of the formation is marked by angular unconformity (with the metasedimentary rocks) or by unconformity (with the granite) basement. The top is a disconformity, usually passing to the Silveirinha dos Figos Formation or to the Falagueira Formation (by erosion in the Nisa area); locally (e.g., at Murracha, in contact with the Ponsul fault), it is by angular discordance with the Torre Formation.

The Cabeço do Infante Formation comprises mainly sandy-conglomeratic facies. The formation includes a lower and an upper member, each one generally displaying a fining-upwards macro-sequence followed by a thickening-upwards macro-sequence. The conglomeratic facies are located mainly in the Monforte da Beira area. The unit corresponds to episodic alluvial sedimentation, with periods of rapid aggradation and erosion by small streams. During intervals of no deposition, paleosoils developed. In this NE sector of the Lower Tejo Basin, low-gradient alluvial fans drained towards the north-west and west. Ordovician quartzite crests, forming important NW–SE paleoreliefs, conditioned the drainage. The lower member appears to have been deposited mainly by unconfined flows (sheet floods), but the upper member architecture is indicative of alluvial sedimentation dominated by braided streams; this diference could have resulted from a change in climate to more arid conditions during the Middle to Late Eocene.

5.6.2 Lower to Upper Miocene

The second major stage of sedimentary infill of the Lower Tejo Basin is represented by allostratigraphic units UBS9 and UBS10. In the NE sector (proximal) of the basin, this record is represented by the *Silveirinha dos Figos Formation* (Cunha 1992, 1996).

The *Silveirinha dos Figos Formation* (Figs. 18, 19, 21, 22, 23, 24, 25 and 26) comprises alternations of thick sandy and silty beds. The sands have an orange colour but the silts are greyish-green, with typical purple-red or orange spots. The sediments are friable and generally poorly sorted. The composition of the sand fraction is mainly feldspar and hyaline quartz grains. Larger clasts consist predominantly of quartzite and quartz pebbles, generally ∼5 cm in diameter and always <10 cm. Channel geometries 100 m wide are typical; massive structures and through cross-bedding are displayed by the sand beds. The unit, with a tabular geometry, is ∼100 m thick. The attitude is usually horizontal, but can be almost vertical at reverse faults. It is deeply dissected by the drainage network, and the main outcrops are located at Sarzedas, Vila Velha de Ródão and south of Idanha-a-Nova.

The formation disconformably overlies the Cabeço do Infante Formation or, more rarely, by angular unconformity, the metamorphic basement. The Silveirinha dos Figos Formation is overlain by the Torre Formation, by the Monfortinho Formation or by Quaternary deposits. In the Sarzedas area, the top of the formation is a disconformity at 400 m of altitude that corresponds to the Lardosa planation surface, formed on the Castelo Branco granites (Castelo Branco surface; Birot 1949), with residual reliefs such as inselbergs to the north-east (Ribeiro 1942).

Vertical and lateral facies changes in the formation are not particularly significant and can be interpreted as resulting from the lateral migration of a belt of active channels over the floodplain. The sedimentary pattern resulted from a sandy, braided depositional system, which drained from the NE to the SW in the Lower Tejo Basin. Depositional and compositional data suggest that the source area comprised primarily the batholiths of Castelo Branco—Monsanto, Idanha and the metasediments located between them. Compared to the previous infilling stage represented by the Cabeço do Infante Formation, there was a change in the paleocurrents and geographical position of the source area, and also better regional drainage. The sedimentary record of this Miocene stage is identifiable throughout the Lower Tejo Basin (Carvalho 1968; Antunes in Ribeiro et al. 1979; Cunha 1992; Barbosa 1995) and corresponds to the Alcoentre Formation in the Ribatejo area. The significant thickness and the vast original extent of this feldspathic unit implies a sustained history of erosion of granitic rocks, ultimately producing a planation surface.

Some layers of the Silveirinha dos Figos Formation show bioturbation or fossil root traces. At Vila Velha de Ródão, silicified trunks were probably provided by this unit, similar to those found at Ponte de Sor. The trunks correspond to *Annonoxylon teixeirae*, an angiosperm tree of which present-day equivalents are typical of tropical/subtropical environments, and which are similar to others obtained from equivalent lithostratigraphic units. This Miocene tree is indicative of a warmer and wetter climate than the present one (Pais 1973; Teixeira and Pais 1976).

At Plasencia (Caceres, Spain), not far east of Castelo Branco, a red to orange conglomeratic and silty unit has provided fossil remains of *Hispanotherium matritensis* (Hernandez-Pacheco and Crusafont 1960), typical of zone MN5 (middle Aragonian, Langhian). Fossils of this steppe regime adapted to xerophytic environments, including rhinoceros, are also abundant in Lisbon in depositional sequence L1, Div Vb (Antunes 1979a; Antunes et al. 1999b), and are also represented in Quintanelas and Amor (Cenozoic Mondego Basin) (Antunes and Ginsburg 1983; Antunes and Mein 1981).

5.6.3 Uppermost Miocene to Gelasian

Overlying the Silveirinha dos Figos Formation, alluvial fan formations occur along tectonic piedmonts related to the uplift of the Portuguese Central Cordillera (Murracha Group, Uppermost Miocene to Pliocene). The thickness of these alluvial deposits and the grain size decreases rapidly downstream. It includes two fining-upwards megasequences (corresponding to the Torre and Monfortinho formations) followed by a coarsening upwards megasequence (Falagueira Formation) constituting the allostratigraphic units UBS11, UBS12 and UBS13 (Figs. 18, 19, 25 and 26) (Cunha and Reis 1992; Cunha 1996, 2000).

The *Murracha Group* records the sedimentary response to the phases of crustal uplift in central Portugal that had begun by the middle Tortonian (Upper Miocene) (Figs. 18 and 19). The Murracha Group disconformably overlies the Silveirinha Figos Formation but, locally, contacts by angular unconformity with the Cabeço do Infante Formation or the Paleozoic basement. The uppermost agradational surface of the Group is the culmination surface of the Neogene sedimentary infill. Fluvial incision during the Quaternary is responsible for the genesis of terrace staircases. The unit displays a downstream decrease in grain size and thickness, typical of alluvial fan deposition. From the bottom to the top of the sedimentary infill of this group, the Torre, Monfortinho and Falagueira formations have been defined.

The *Torre Formation* (Figs. 18 and 19) comprises an alternation of gray-green gravels and micaceous clayey sands at the base passing to a predominance of brown-yellowish or green silts at the top. The sediments are poorly sorted and have an abundant clay matrix. In general the sediments are friable, but locally they can be cemented by silica. Clasts of phyllite and metagraywacke predominate over those of white quartz and quartzite. The sands are generally sub-arkosic and micaceous. The clay mineral association comprises smectite with some illite.

Along the Ponsul and Sarzedas fault scarps, the formation constitutes endorheic alluvial fan deposits, reaching 100 m thick, probably representing the upper Tortonian to Messinian. In the piedmont, reddish brown conglomeratic facies occur, locally alternating with green or reddish orange sandy silt layers (Vale

Bonito Member). Distally (towards the SE), the thickness of these facies decreases and grayish or yellowish green sandy micaceous silts are dominant (*Sarzedas Member*).

Locally, the unit displays folding near reverse NE–SW faults, which indicates a later tectonic phase. The lower and upper limits of the formation are regional sedimentary descontinuities, and the sedimentary record also corresponds to allostratigraphic unit UBS 11 (Cunha 1996, 2000).

The *Monfortinho Formation* (Figs. 18 and 19) consists of red alluvial fan conglomeratic facies (*Piçarra Vermelha Member*) grading upwards and distally to sandy lutites (*Cantareira Member*). Larger clasts are dominated by white quartz, quartzite, phyllite and metagraywacke. Clasts become larger and more angular nearer the paleo-scarps. In the sand fraction, white quartz dominates over hyaline quartz, but some phyllite and feldspar grains are also present. The clay mineral association shows illite and kaolinite in similar proportions. This formation crops out in the regions of Sarzedas and Monfortinho, in small remains dissected by Quaternary fluvial erosion. At the Murracha and Murrachinha buttes the unit is 80 m thick, but reaches 130 m at the Monfortinho quartzitic crest. The stratification is sub-horizontal. It overlies, by low angle disconformity, the Torre Formation, the Silveirinha dos Figos Formation or the Variscan basement (sometimes fossilizing tectonic scarps). Locally, the unit has erosive contacts with the Falagueira Formation or Quaternary deposits. This sedimentary record also corresponds to allostratigraphic unit UBS12 (Messinian to Zanclean). It documents deposition in endorheic alluvial fans, most probably under a temperate dry climate with strong seasonal contrasts.

The *Falagueira Formation* comprises ochre, sometimes whitish or reddish, alluvial fan conglomeratic facies and culminates the Cenozoic sedimentary infill (Figs. 18 and 19). It can be ascribed to allostratigraphic unit USB 13 (uppermost Zanclean to Gelasian). It reaches 107 m in thickness and has a horizontal attitude. Near the Moradal and Penha Garcia quartzitic ridges, the deposits are thick and very heterometric, but further south (distally), they are less coarse and more regularly organized. The sediments are poorly sorted conglomerates, with a matrix of silt to coarse sand. Their composition is dominated by clasts of quartzite and of quartz, with the clay mineral association being rich in kaolinite with accessory illite.

The formation disconformably overlies the Monfortinho and the Cabeço do Infante formations (e.g., Magarefe butte) and, in some cases, the basement (e.g., Falagueira-Malpica). Locally, some fluvial incision into the older units is evident. The altitude of the surface of the unit increases with distance from the Tejo River or with increasing proximity to the Portuguese Central Cordillera, where it is preserved as buttes (e.g., Murracha, Murrachinha and Pedras Ninhas, reaching 577 m in altitude) or as mesas (Sarzedas is 1600 m long, 447 m in altitude).

The formation indicates a coarsening-upwards macro-sequence, reflecting prograding alluvial fans and tributaries of braided rivers draining into the Atlantic, ancestral components of the present-day drainage system (Cunha et al. 1993). In proximal sectors, heavy rains and intense weathering of clay shales would have favoured landslides and debris-flows (Ribeiro 1942) grading downstream to torrential streams. The abundance of quartzite blocks would have resulted from the erosion of rejuvenated quartzite ridges.

In the area of Sarzedas, the remains of the Falagueira Formation testify to an alluvial fan controlled by the activity of the Pomar fault (NNE–SSW direction, with uplift of the western block). At the Magarefe butte, just at the foot of the Moradal quartzite ridge (top at 815 m), the unit is thicker (>100 m) and the blocks of quartzite can reach 2 m across. At Sarzedas and Cantareira villages, the formation is ~10 m thick and the blocks are larger at the top of the unit, reaching 60 cm. The percentage of clasts of phylite or metagraywacke is higher at the Sarzedas mesa, especially at the base; channel structures indicates local paleocurrents towards the ESE.

Between Nisa and Monforte da Beira, the facies correspond to deposits of a gravelly river with longitudinal bars. A large river was oriented WSW–ENE to W–E, an ancestor of the present-day Tejo River, which had already captured the drainage of the High Tejo Basin (Spain). At Monfortinho, a coeval alluvial fan fed a gravel braided river (the ancestral Erges River), with drainage to the south-west, crossing the quartzite ridge.

In summary, the Falagueira Formation was deposited in a context of alluvial fans and gravelly braided rivers draining into the Atlantic, ancestors of the present-day rivers (Cunha et al. 1993). The deposits and their substrate exhibit weathering features indicative of leaching conditions, with important kaolinization and hydromorphization. The spatial development of rivers, the predominance of clasts highly resistant to weathering, and the significant predominance of kaolinite over illite, together suggest the persistence of water in the landscape under a hot and humid climate.

The Falagueira Formation, defined in the NE sector of the Lower Tejo Basin (LTB), is correlative of the Almeirim and Ulme Formations (Barbosa and Reis 1989) defined in the middle sector of the LTB.

Younger deposits (terraces, colluviums and alluviums) are distinguished by their lower topographic position, already related to the stage of incision, and are usually more heterogeneous.

5.7 Paleogeographic Evolution

The first paleogeographic reconstructions of the distal sector of the LTB were presented by Antunes (in Ribeiro et al. 1979). New data, including those provided by the study's survey of Belverde (Pais et al. 2003; Legoinha et al. 2004) and the characterization of allostratigraphic units in the intermediate and proximal sectors have allowed the establishment of more detailed palaeogeographic maps (Figs. 20, 21, 22, 23, 24, 25, 26 and 27).

During the Eocene (Middle?), when started the opening of the Lower Tejo Basin deposition began in the form of extensive alluvial fans of siliciclastic

Fig. 20 Paleogeographic reconstruction of the Lower Tejo Basin for the Lutetian (in part from Cunha et al. 2009)

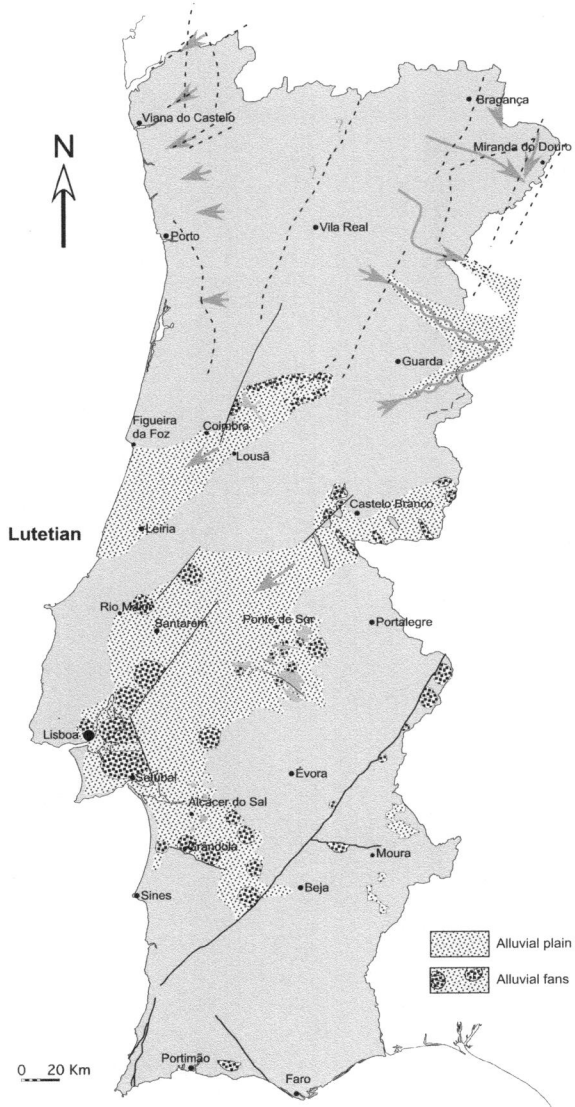

heterometric material near slopes of the surrounding relieves. The regime was endorheic during the Paleogene, and there was a favourable climate for the formation of carbonate crusts and calcretes.

In the first Miocene transgression, the sea penetrated the Setúbal Peninsula, with the gulf region reaching Lisboa. Seismic profiles and data provided by the Belverde borehole suggest that the marine entry was made from the south. The warm waters, and possibly a previously-formed structural high, favoured the

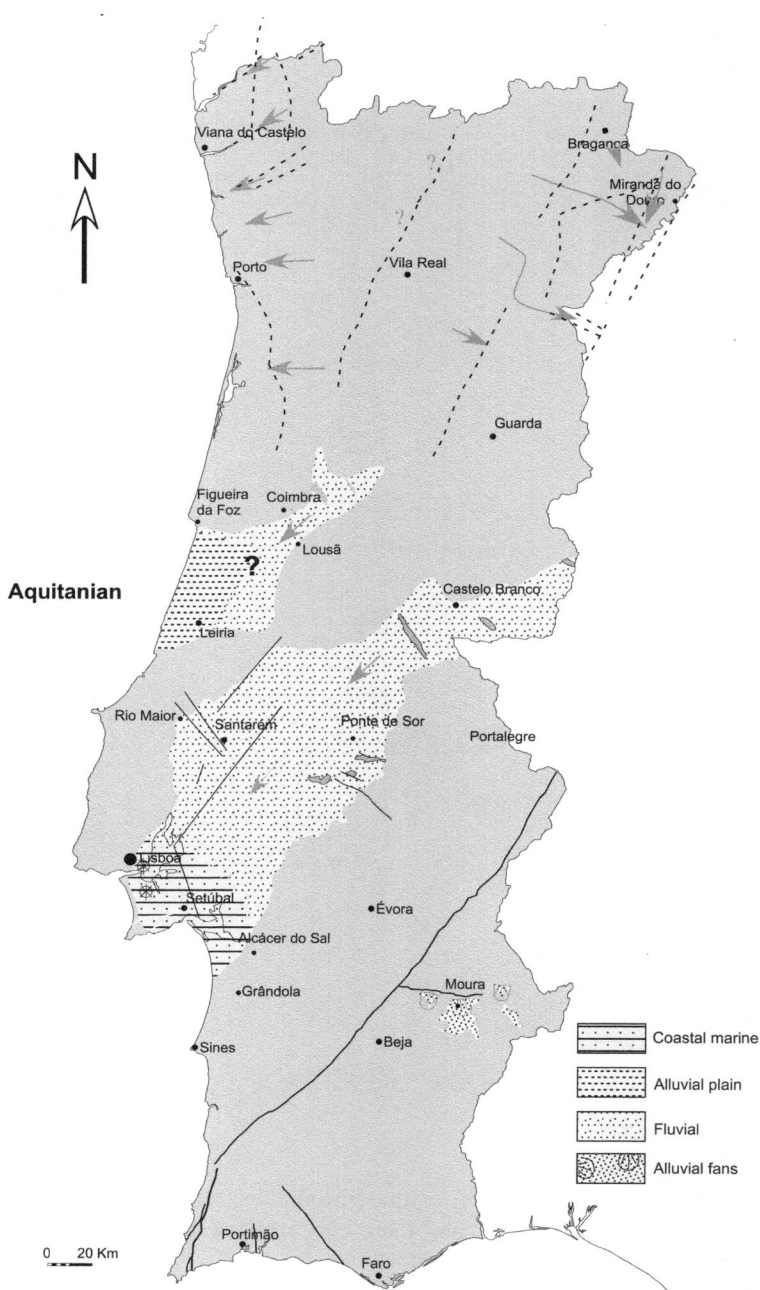

Fig. 21 Paleogeographic reconstruction of the Lower Tejo Basin for the Aquitanian (in part from Cunha et al. 2009)

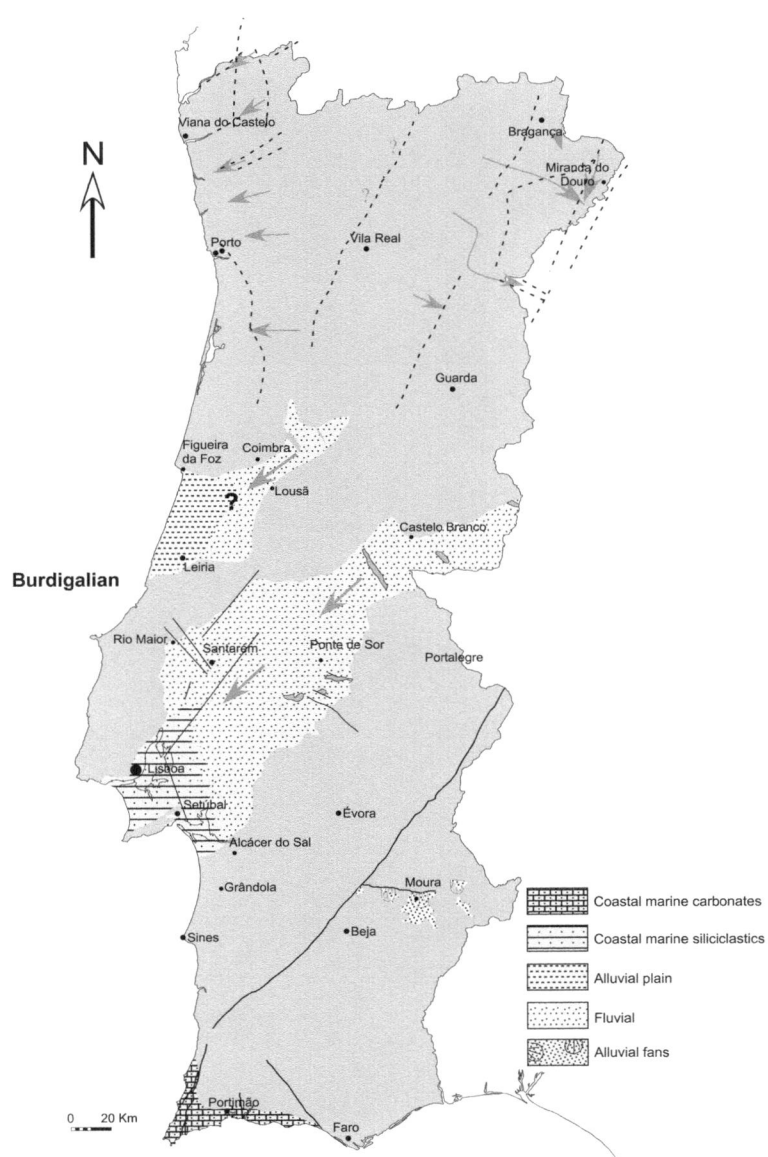

Fig. 22 Paleogeographic reconstruction of the Lower Tejo Basin for the Burdigalian (in part from Cunha et al. 2009)

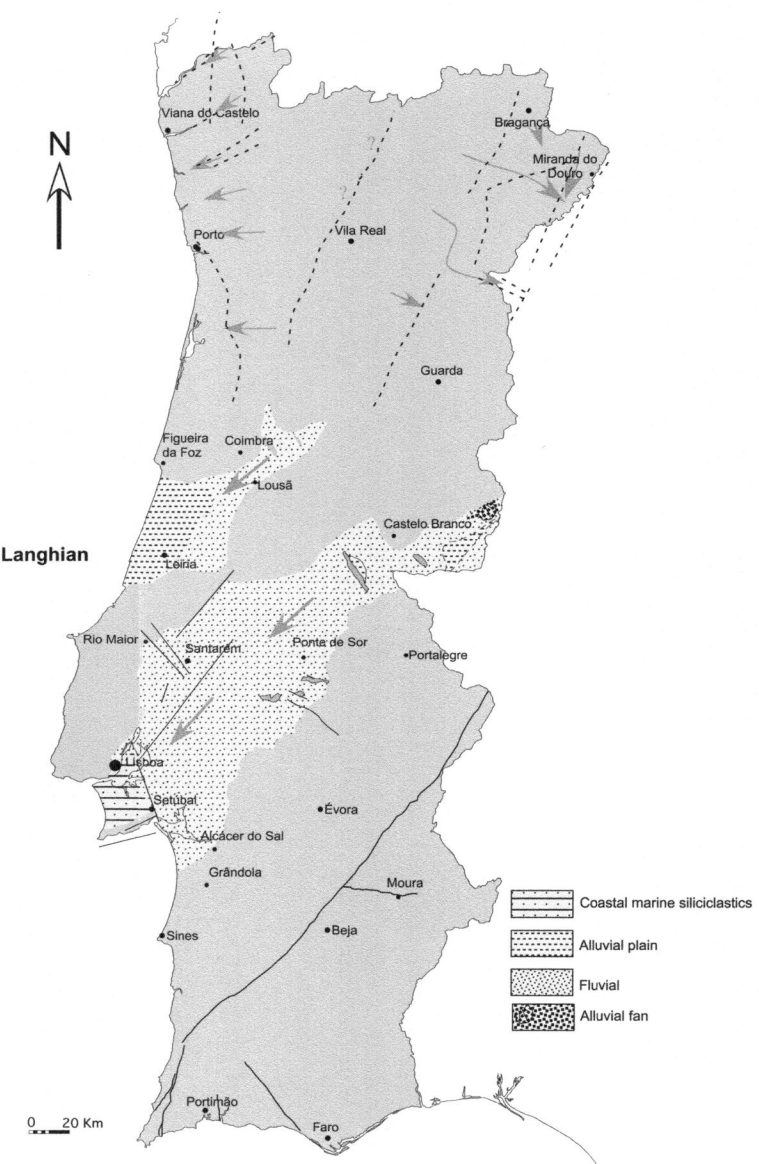

Fig. 23 Paleogeographic reconstruction of the Lower Tejo Basin for the early Langhian (in part from Cunha et al. 2009)

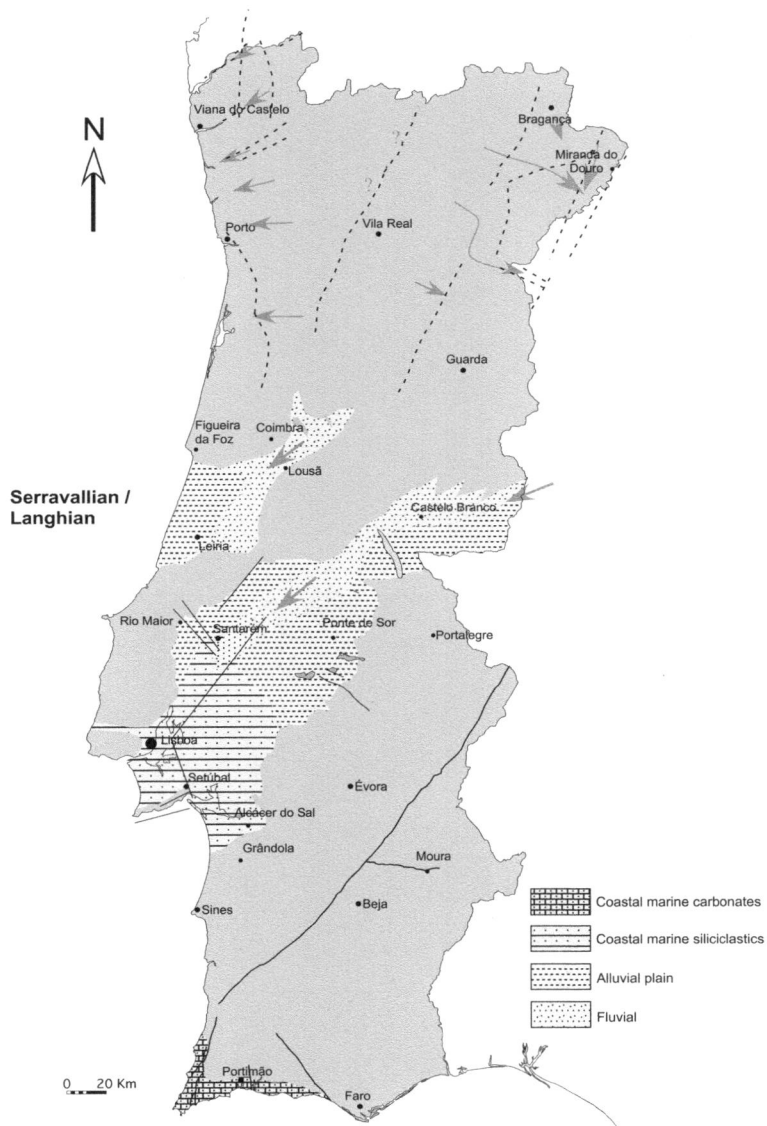

Fig. 24 Paleogeographic reconstruction of the Lower Tejo Basin for the late Langhian and Serravallian (in part from Cunha et al. 2009)

development of coral reefs by creating a barrier oriented N–S. The western Atlantic opening took place in the early Burdigalian (early depositional sequence B1). A high seabed oriented N–S and approximately coincident with the existing shoreline protected the inner sector of the gulf of the Setúbal Peninsula.

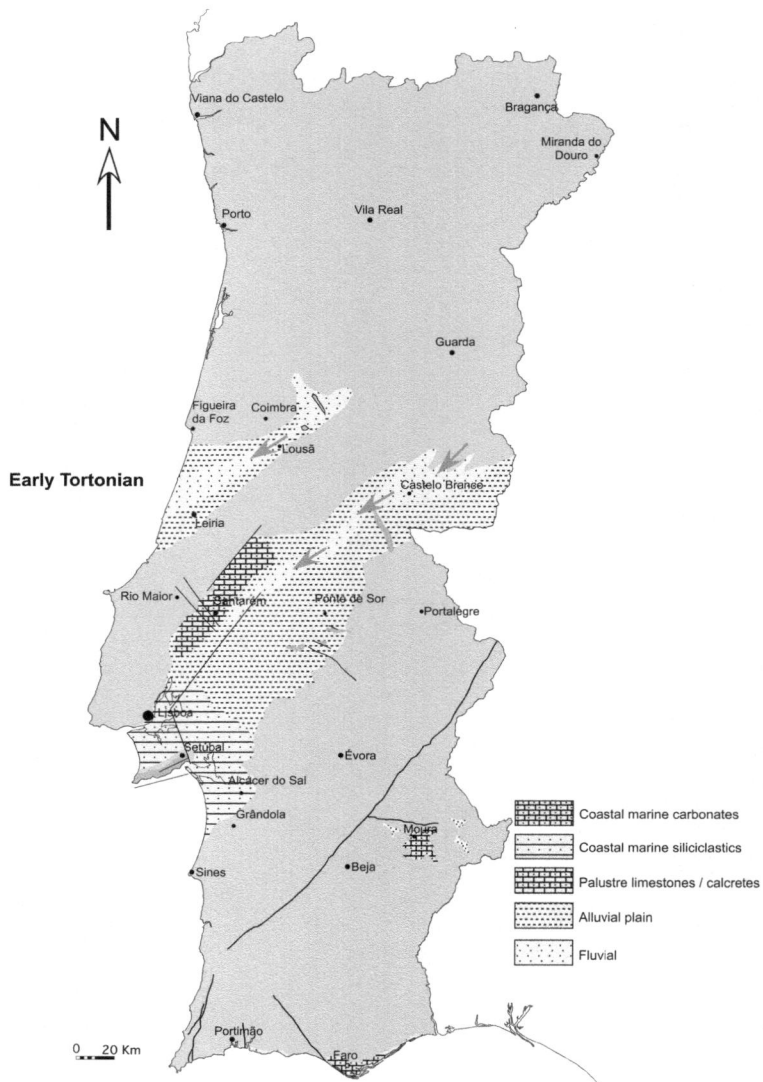

Fig. 25 Paleogeographic reconstruction of the Lower Tejo Basin for the early Tortonian (in part from Cunha et al. 2009)

Subsidence allowed the deposition of >1000 m of Neogene sediments, while shallower parts on the sedimentary thickness reached only 200 m, with stratigraphic gaps being recognized here between 17 and 14.8 Ma and between 12.7 and 11.6 Ma (Legoinha 2008) However, the lower limit of this later hiatus is positioned at 13.5 Ma using the ICS 2009 time scale (Fig. 11).

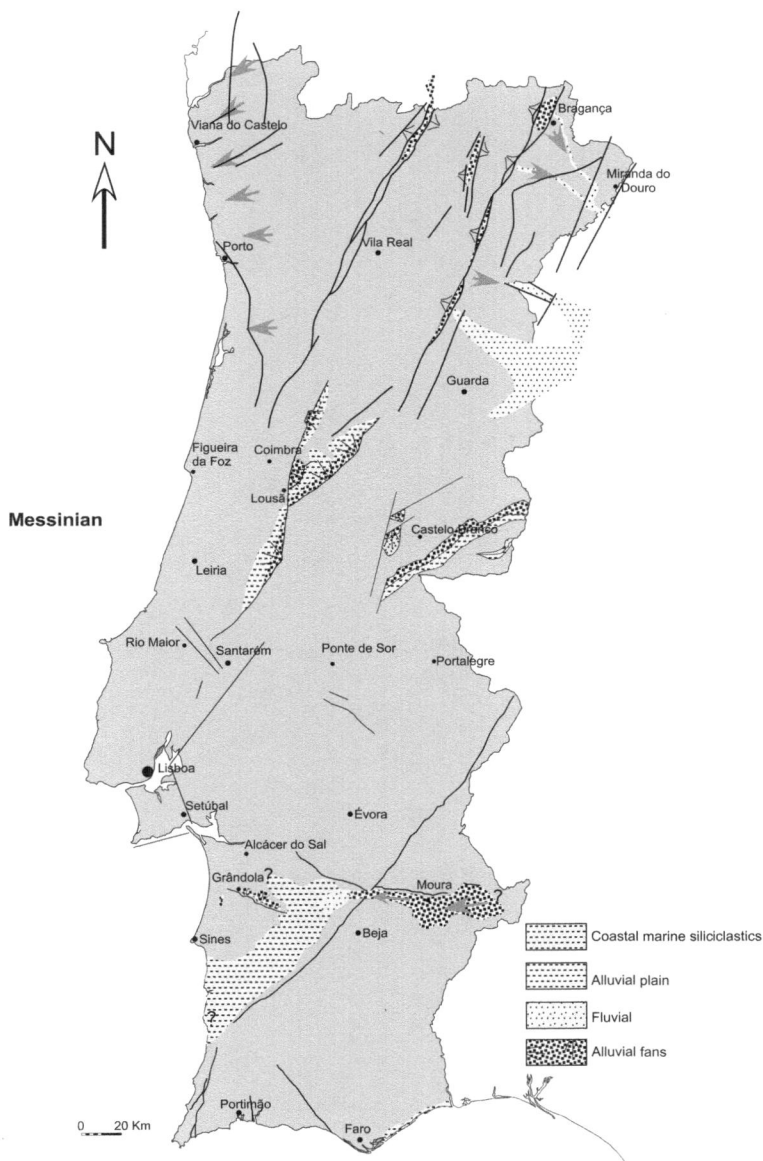

N

Messinian

Viana do Castelo
Bragança
Miranda do Douro
Porto
Vila Real
Guarda
Figueira da Foz
Coimbra
Lousã
Castelo Branco
Leiria
Rio Maior
Santarém
Ponte de Sor
Portalegre
Lisboa
Setúbal
Évora
Alcácer do Sal
Grândola
Moura
Sines
Beja
Portimão
Faro

0   20 Km

Coastal marine siliciclastics
Alluvial plain
Fluvial
Alluvial fans

Fig. 26 Paleogeographic reconstruction of the Lower Tejo Basin for the late Messinian (in part from Cunha et al. 2009)

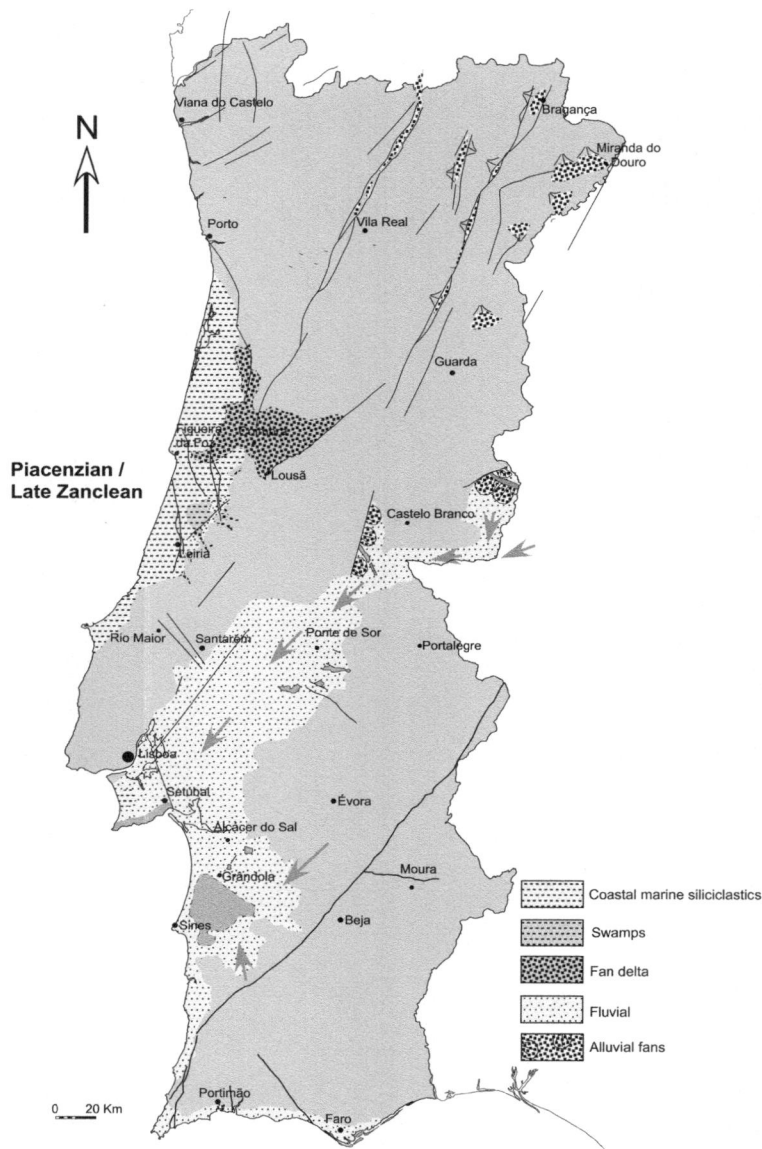

Fig. 27 Paleogeographic reconstruction of the Lower Tejo Basin for the late Zanclean to Piacenzian (in part from Cunha et al. 2009)

During eustatic high levels (mostly in the middle Burdigalian and late Langhian), the brackish waters extended ∼100 km into the basin. The Arrábida chain, freshly uplifted, was at this time an island, and the same situation is inferred for Sintra mountain (Figs. 19, 22 and 23).

During the Early and Middle Miocene in the intermediate sector, the Tejo River rambled over an extensive floodplain, depositing the siliciclastic Alcoentre Formation. In the early Tortonian, an extensive swampy area, with lakes in Ribatejo, was installed. The limestone Almoster Formation was accumulated, as well as clays, calcretes and carbonate concretions formed as the Tomar Formation (Figs. 19, 24 and 25).

In the Pliocene, widespread progradation occurred. Fluvial sands entered the area of the Setúbal Peninsula (Ulme and Santa Marta Formations). Gravels prograded to the SW (Almeirim and Falagueira Formations) from the most proximal areas. During this interval, climatic conditions were favourable for the generation of ferruginous concretions and crusts, providing a red–orange colour to the sediments (Fig. 27).

During the Pleistocene, river terraces, alluvium and aeolian sand were generated (e.g., Cunha et al. 2005, 2008a, b; Martins and Cunha 2009; Martins et al. 2009a, b, 2010a, b).

6 The Guadiana and Moura Basins

6.1 General Aspects

The Cenozoic Guadiana Basin covers a large area lying mostly in Spain. The basin comprises two main remnants, separated by a tectonically controlled granitic threshold located in the Merida region. These two areas are known as "Vegas Bajas del Guadiana", between Portugal and Merida, and "Vegas Altas del Guadiana", situated upstream (e.g., Hernández Pacheco 1960). The sedimentary basin fill includes lacustrine, fluvial and alluvial fan deposits, unconformably overlying metamorphic and igneous basement rocks from the Iberian Massif. This sedimentary record reflects not only climatic and paleoenvironmental changes, but also important paleogeographic modifications, in which the general Cenozoic drainage pattern, oriented towards the west, was temporarily endorheic (Garzón 2005). These regional paleodrainage shifts were probably related to alpine tectonic activity, in which local subsidence and the reactivation of a NE–SW fault system (e.g., the Messejana-Ávila fault zone) played an important role (Moya Palomares et al. 2000). In the Badajoz-Merida region (Tierra de Barros), the following lithostratigraphic division is generally accepted (Villalobos et al. 1988):

(a) Quaternary fluvial deposits;
(b) Plio-Quaternary raña deposits;

(c) Plio-Quaternary fluvial deposits;
(d) Cenozoic Upper Unit (Facies Almendralejo, Facies Badajoz and Caleño); and
(e) Cenozoic Lower Unit (Facies Lobón).

In Portugal, only the western limit of the Guadiana basin crops out, in Elvas, Juromenha, Campo Maior and Redondo regions (Alto Alentejo). These sub-basins are interpreted as small pull-apart depressions, formed along the Messejana-Ávila fault zone by left lateral strike-slip movement on releasing bends or on strike-slip duplex segments (Brum da Silveira 1990; Villamor 2002). The sediment fill includes Miocene sequences of reddish clay deposits, coarse to fine sandstones and conglomerates, with a calcic horizon of pedogenic origin developed on top of the sequence, and Pleistocene conglomeratic facies associations related to near-source alluvial fans, derived from fault scarp degradation.

The Cenozoic Moura Basin area covers mostly the Moura-Marmelar depression located north-east of Beja (Fig. 1). Several isolated and smaller sub-basins (e.g., the Amareleja, Safara, Oriola) are also included in the Moura Basin framework, due to an equivalent stratigraphy, similar sedimentary characteristics and the same morphotectonic context.

The Moura-Marmelar basin is bordered by the Vidigueira-Moura fault zone (VMFZ), a 65-km-long, E-W trending, N-dipping reverse left-lateral late Variscan structure, which has been reactivated during the Paleogene, Miocene and Pleistocene with reverse and right-lateral strike-slip movement. Cenozoic fault reactivation is denoted by geomorphological, stratigraphic, and structural data, in response to a NW–SE-trending compressive stress (Brum da Silveira et al. 2009). The lithostratigraphic model comprises four unconformity bounded depositional sequences related in the main to alluvial fan systems (conglomerates, sandstones), palustrine systems (limestones) and fluvial systems (sandstones, conglomerates, claystones), and two major episodes of pedogenic carbonate development (calcretes) (Brum da Silveira 1990, 2002).

Facies recurrence, the scarcity or absence of fossils, and the lack of geochronometric dating are considered the major difficulties in establishing stratigraphic correlations between sub-basins and evaluating the evolution of the Moura Basin in a regional context.

6.2 The Guadiana Basin

6.2.1 Oligocene/Miocene

Cenozoic Lower Unit-Facies Lobón

The Cenozoic Lower Unit (*Facies Lobón* or *Arcillas Rojas de Lobón*) comprises a monotonous sequence (80 m thick) of reddish clay deposits and fine clayey sandstones, related to a lacustrine depositional environment or distal alluvial fan system. This unit unconformably covers a thick alteration profile developed on Precambrian and Paleozoic basement (Rodríguez Vidal et al. 1988).

6.2.2 Miocene

Cenozoic Upper Unit

Facies Almedralejo/Facies Badajoz

 The Cenozoic Upper Unit lies disconformably over the Facies Lobón deposits and by a non-conformity over Precambrian and Paleozoic basement rocks. It comprises two fluvial facies associations that are laterally interconnected (Facies Almendralejo and Facies Badajoz), suggesting a braided river system that drained the Cenozoic basin to the west (Villalobos et al. 1988; Villalobos and Jorquera 1998). In proximal sectors, located in the eastern part of the basin, Facies Almendralejo consists of conglomerates, microconglomerates and sandstones, sometimes with cross-bedding and a carbonated cement, revealing channelized architecture and coarsening-upward sequences. In distal sectors, located in the western part of the basin, Facies Badajoz consists of sandstones, fine sand and silt deposits, in channelized fining-upward sequences (Duran et al. 2005).

Facies Caleño

Culminating the Cenozoic Upper Unit, a carbonated deposit known locally as the "caleño" was interpreted initially as a lacustrine deposit formed in a semiarid region (Armenteros et al. 1986). Presently Facies Caleño is interpreted as a carbonate paleosoil, comprising calcic horizons formed by the accumulation of pedogenic carbonates (calcretes), developed in an arid climate characterized by a dry season and occasional heavy rains.

6.2.3 Plio-Quaternary

Rañas Unit

The "Rañas Unit" comprises poorly sorted conglomerates with a red, sandy-clay matrix, with sub-rounded quartzite, quartz and shale clasts (up to 20 cm). These deposits are associated with alluvial fans systems. The unit rests unconformably over Iberian Massif basement rocks and the Tertiary Lower and Upper Units.

6.3 The Moura Basin

6.3.1 Paleogene-Miocene

Depositional Sequence DS1

Casa Branca Lower Unit

The Casa Branca Lower Unit (Fig. 28) includes, in proximal areas, medium- to coarse-grained siliciclastic sandstones (Sc to Sm; Miall 1996) and conglomerates (Gms and Gm) composed of angular quartz pebbles with some schist, meta-volcanic and metalidite clasts; in distal areas it comprises medium- to fine-grained sandstones (Sm) and siltites (Fm and Fsc). The unit has a maximum thickness of 10–15 m, resting unconformably on a relatively smooth paleotopography cut into the Proterozoic and Paleozoic basement.

The sediments are often strongly cemented, showing a pedogenic calcic horizon and, sometimes, a pedogenic silcrete horizon. The morphotectonic context, sedimentary characteristics and depositional architecture suggest the occurrence of syntectonic alluvial fan deposits as a sedimentary response to reactivation of the Vidigueira-Moura Fault Zone, originating from an elevated source area to the north; their age is uncertain, though they most likely range from the Paleogene to Early Miocene (Brum da Silveira 1990, 2002).

6.3.2 Upper Miocene

Depositional Sequence DS2

Marmelar-Moura Intermediate Unit

In the western sector of the Moura-Marmelar basin, DS2 is related to syntectonic coalescent alluvial fans systems from the VMFZ fault scarp (proximal areas) reflecting a Late Miocene reactivation phase on this tectonic structure. In this sector, the Intermediate Unit comprises conglomeratic facies associations (Gms to Gm), constituted predominantly by sub-angular quartz and schist clasts, and medium to coarse-grained sandstones (Sc), with discontinuous impregnations of pedogenic carbonates (powder calcretes).

In the eastern sector of Moura-Marmelar basin (Moura), and also in Amareleja and Safara sub-basins, DS2 is related to a fluvial depositional system, a pre-Guadiana river draining to the west (Figs. 1, 26 and 28); these fluvial deposits include siliciclastic, medium- to fine-grained sandstones (Sm) and siltites (Fm and Fsc), grading upwards to coarse sandstones (Sc and Sp) with some interbedded conglomerates (Gm and Gms), composed predominantly of subangular to sub-rounded quartz and schist pebbles. Also in the eastern area, and culminating the DS2, palustrine carbonate deposits occur, related to an endorheic drainage often showing a lateral interconnection with pedogenic carbonate accumulations (calcretes).

This unit has a maximum thickness of 30 m, and rests disconformably over the DS1 deposits or in non-conformity over the eroded Variscan basement. A Late Miocene age is considered likely for this unit.

6.3.3 Lower Pleistocene

Depositional Sequence DS3

Mesas Unit

The Mesas Unit (Fig. 28) attains a maximum thickness of approximately 25 m and comprises reddish, poorly sorted conglomerates (Gms and Gm) with sub-angular and sub-rounded quartz and schist pebbles, and some medium- to coarse-grained sandstones (Sc). These sediments usually occupy summit positions in the present morphology, and culminate the sedimentary infill, except in a depressed area (Vale da Serra graben) where they are overlain by the DS4 sediments. They paraconformably overlie the DS2 deposits, or rest locally in angular unconformity over the DS2 and/or DS1 deposits, or non-conformably over the Variscan basement rocks. The nature, lateral facies changes, and depositional architecture of these sediments indicate alluvial fan systems derived from the VMFZ fault scarp located to the north. However, in some areas, far from the fault scarp, correlative sediments support a non-tectonic origin, deriving from residual reliefs of Cambrian quartzites and dolomitic limestones, thus pointing to semi-arid climatic conditions coeval with the tectonic activity (Brum da Silveira 1990, 2002).

6.3.4 Pleistocene

Depositional Sequence DS4

Vale da Serra Unit

The Vale da Serra Unit (Fig. 28) has a thickness of 20-30 m and comprises siltites (Fsc and Fm), fine- to coarse-grained siliciclastic sandstones (Sm and Sc) and conglomerates (Gm and Gp) composed of angular to subrounded quartz and schist pebbles. At the top, it shows a ferruginous paleosol with strongly developed rubification and cementation. These sediments disconformably overlie all the previous depositional sequences (DS1, DS2 and DS3), and are disconformably overlain by Pleistocene fluvial terrace deposits. DS4 facies associations indicate streamflow depositional processes in coalescent alluvial fans feeding Vale da Serra graben. This depositional sequence is interpreted as the sedimentary response to a tectonic event of fault reactivation and corresponding morphogenesis. A Pleistocene age is attributed to these deposits, on the basis of their topographic and

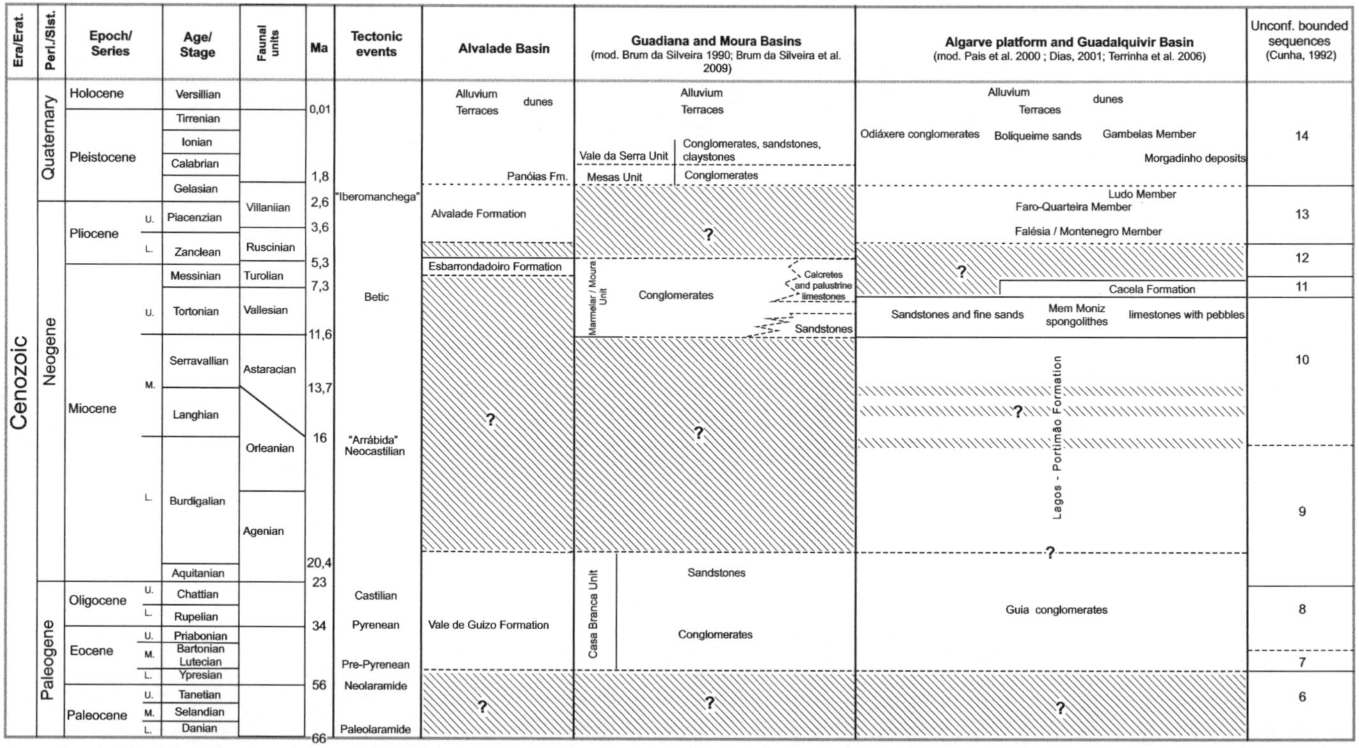

Fig. 28 Correlations between units from the Cenozoic Alvalade, Guadiana and Moura basins, and deposits of the Algarve platform and the western Guadalquivir Basin (Portugal), including the unconformity-bounded sequences (updated, Cunha et al. 2009)

stratigraphic basin emplacement (Brum da Silveira 1990; Brum da Silveira et al. 2009).

7 The Alvalade Basin

7.1 General Aspects

South of the Lower Tejo Basin, an area that largely corresponds to the Sado hydrographic basin, the Alvalade Basin, developed (Antunes and Mein 1989). This basin is separated from the Lower Tejo Basin by a horst formed by Paleozoic rocks (Senhor das Chagas-Valverde), which has been crossed by the Sado River only during the Quaternary (Figs. 1, 2 and 28).

For many years, the separate identity of this basin was not recognized and it was considered integrated with the Lower Tejo Basin as the Tejo-Sado Basin (Carvalho et al. 1985). However, the horst of Senhor das Chagas-Valverde has acted as a paleogeographic barrier between the two. Precise dating of marine deposits, superimposed on the mainland units generally attributed to the Paleogene (Vale de Guizo Formation) has shown that to the north and west of the horst ages are ascribed to the late Serravallian-early Tortonian, while to the south, ages are ascribed to the Messinian (Antunes and Mein 1989).

Thus, the filling of the two depressions after the Miocene, due to the lifting of the horst of the Senhor das Chagas-Valverde, have different and independent histories and should be seen as distinct morphological units: the areas north of the barrier form part of the Lower Tejo Basin, and those to the south constitute an independent basin (the Alvalade Basin).

The genesis and structure of this basin are associated with one of the first compressive Alpine tectonic phases, related to the convergence of the Eurasian and African plates during the Cenozoic.

7.2 Paleogene

The *Vale do Guizo Formation* (Fig. 28) (Antunes 1983; Gonçalves and Antunes 1992) corresponds to the initial stage of basin infilling. It rests directly on Paleozoic metasedimentary basement. The formation is represented mainly by reddish, coarse sandy-gravelly deposits, with abundant calcretes or smectite clays and groundwater dolocretes or paligorskite at the top (Azevêdo and Pimentel 1995; Pimentel 1997, 1998a, 2002). The deposits have massive and graded structures, without tractive macroforms, indicating the predominance of debris flows and mud-flows in fining-upwards macro-sequences. The unit is inferred to represent alluvial fans generated under sub-arid climate and which developed in response to tectonic events. The palaeogeographic development of the Torrão and Messejana sub-basins are related to the two main fault scarps in the Alvalade basin (Pimentel 1997, 1998a).

The Vale do Guizo Formation is correlated with other deposits attributed to the Paleogene, such as at Vale Furado and Côja (Mondego Basin), dating biostratigraphically from the Eocene, as well as the Benfica, Monsanto and Cabeço do Infante Formations (from the Lower Tejo Basin) (Antunes 1983, 1996b; Cunha 1992).

These deposits all show similar tectonic, climatic and geomorphological contexts; in particular, they represent a response to the activation in the Paleogene of tardi-Variscan faults, leading to subsidence of intracontinental basins with probable N–S compression and development of alluvial fans. The features allow lithostratigraphic and tectono-sedimentary correlation to be made with the first sequence of the Paleogene deposits from central Portugal, attributable to the interval between the Middle Eocene and the lower Oligocene (Cunha 1992).

7.3 Upper Miocene

Overlying the Vale do Guizo Formation is the *Esbarrondadoiro Formation* (Fig. 28). The formation includes various facies, predominantly yellowish micaceous fine sand, with micro-conglomerates and lutite intercalations (rich in smectite). The layers have a large-scale tabular geometry, with horizontal stratification and positive gradation (Pimentel 1998b); outcrops are extensively distributed.

The formation corresponds essentially to a fossiliferous marine unit. It contains abundant fauna including oysters, Pectinidae, selachian teeth, chelonians, remains of whales and small and large mammals attributable to zone MN13 ($\sim$6 Ma, Messinian) (Teixeira 1952; Antunes and Mein 1995; Balbino 1995; Antunes et al. 1999a; Antunes and Balbino 2004, 2006). Recent $^{87}Sr/^{86}Sr$ age determinations of oysters from Alfundão (Ferreira do Alentejo) gave 6.4 (+0.4,−0.3) and 6.0 (+0.3,− 0.2) Ma (isotopic analysis by MacArthur, Univ. College, London).

The small mammal fauna is very similar to the faunas of the terminal Miocene of Spain and North Africa. It indicates conditions warmer than the present and a degree of dryness, though with Mediterranean characteristics. The mammal fauna includes (Antunes and Mein 1989, 1995):

Insectivora

Episoriculus sp.
Galerix iberica

Lagomorpha

Prolagus michauxi
Trischizolagus maritsae

Rodentia

Eliomys truci
Eliomys intermedius
Cricetus barrieri
Ruscinomys lasallei
Blancomys sanzi
Stephanomys dubari
Occitanomys alcalai
Apodemus gudrunae
Paraethomys meini
Paraethomys abaigari
Castillomys margaritae

Equidae

Hipparion sp.

Girafidae

Girafidae ind.

Proboscidae

Cf. *Tetralophodon longirostris*

Cetacea

Misticeti
Odontoceti

The ichthyofauna (selachians) of the Esbarrondadoiro Formation (Balbino 1995; Antunes and Balbino 2004, 2006) is characterized by abundant Odontaspididae, Scyliorhinidae, Hemigaleidae, Carcharhinidae, Rhinobatidae, and Dasyatidae Myliobatidae, including species comparable to current forms: *Carcharhinus* cf. *perezi*, *Dasyatis* gr. *centroura*, *Dasyatis* gr. *gigas*, *Dasyatis pastinaca* or *marmorata*, *Dasyatis* cf. *margaritelli* and *Taeniura* cf. *grabata*. Numerous genera found still live in the Mediterranean and the Atlantic: *Carcharias*, *Isurus*, *Scyliorhinus*, *Carcharhinus*, *Galeorhinus*, *Mustelus*, *Rhizoprionodon*, *Sphyrna*, *Squalus*, *Squatina*, *Rhinobatos*, *Torpedo*, *Raja*, *Dasyatis*, *Taeniura*, *Myliobatis*, *Pteromylaeus*, *Rhinoptera* and *Mobula*.

Typical forms from the Miocene are absent: *Carcharhinus priscus*, *Dasyatis rugosa* and *Aetobatus arcuatus* and current forms appear as the first found in the fossil record (*Carcharhinus perezi*, *Dasyatis pastinaca* or *Dasyatis marmorata*, *Dasyatis margaritelli*, *Taeniura grabato*). The species *Dasyatis* is very abundant and different from the current fauna; *Raja* is rare, in contrast to the present-day.

The sea temperature was cooler than the current values of Cape Verde and Senegal, and would have been similar to the modern Atlantic off the coast of Mauritania and Morocco.

The character of the selachian fauna, including the absence of *Carcharodon carcharias*, the evolved character of *Carcharhinus* and the absence of *Hemipristis* corroborate the pre-Pliocene (Miocene to end of Messinian) age designation provided by small mammals (Antunes and Mein 1989) for the Esbarrondadoiro Formation.

Given the rarity of pelagic elements *Carcharocles megalodon* and *Isurus hastalis*, the environment was likely to be that of a relatively narrow, unconfined gulf. The regional distribution of sedimentary facies also suggests a narrow marine gulf, surrounded by a paralic area. The more marginal areas, where alluvial fans and eventually deltas developed, received coarse material from the regions to the east.

The unit can be correlated, at least in part, to the upper part of the Cacela Formation from the Algarve (Antunes and Mein 1989; Antunes et al. 1996c; Legoinha 2001), and perhaps also to the Rio da Prata sands at Setúbal Peninsula. However, precise age determinations provided by small mammals have shown that the Esbarrondadoiro Formation dates from the late Messinian to early Zanclean (UBS 12), and is younger than the Algarve Cacela Formation (final Tortonian to Messinian) and the higher sea levels dated in the Lower Tejo Basin (middle Tortonian) (Antunes and Mein 1989). This unit represents deposits corresponding to the tardi-Miocene transgression, related possibly to the third-order cycle 3.3 of Haq et al. (1988).

Pimentel (1998b) characterized the *Monte Coelho Formation* along the southern edge of the Alvalde Basin. New field observations suggest that these materials may correspond to a local facies change from the Esbarrondadoiro Formation. The unit consists of coarse conglomerates with quartz clasts and smectite matrix, and presents abundant hydromorphism (reddish to yellowish); sandstone layers are intercalated with lutites and dispersed carbonate and iron concretions (Pimentel 1998a). The facies are mostly massive, with little organization or grading. The unit has been interpreted as corresponding to the reactivation of the Messejana Fault and to the development of small, proximal alluvial fans under semi-arid conditions.

So, the Formation could be correlated with the sedimentary response to the two major tectonic events responsible for the uplift of several high terrains in Portugal and the coeval alluvial deposits possessing the same sedimentary signature (UBS 11 - upper Tortonian to lower Messinian; UBS 12 - upper Messinian to Zanclean). Therefore, a tectono-sedimentary correlation between this formation and the reactivation of the Messejana Fault, with compression oriented NNW–SSE.

7.4 Pliocene to Lower Pleistocene

The *Alvalade Formation* (Fig. 28) consists of medium-grained orange sands arranged in homogeneous metric layers. These deposits have facies, structure and geometry typical of fluvial environments, corresponding to braided rivers evolving vertically from type 2 to types 3 and 9 (Miall 1985) in an autocyclic positive macro-sequence.

The paleogeographic reconstruction shows a general drainage pattern oriented towards the NW, with no significant fault scarp detected. This formation is considered to be a lateral equivalent of other fluvial units, and probably comprising the Late Pliocene (UBS 13 - uppermost Zanclean to Gelasian). These deposits seem to be contemporaneous with the Ibero-Manchega I tectonic phase, which was responsible for the general uplift of Iberia and its tilting to the SW. This situation, coupled with the more humid conditions of the Piacenzian, encouraged the development of exorheic drainage.

Towards the top, the unit becomes coarser, consisting essentially of layers of gravelly sand.

7.5 Pleistocene

The *Panóias Formation* (Fig. 28) consists of dark red, very coarse gravels with clay matrix (illite and kaolinite), intensely ferruginized by pedogenic processes. The sedimentary structures and textural characteristics of these deposits indicate debris flows associated with large alluvial fans. However, the source area was not the fault scarps bordering the basin, but the Caldeirão Mountain, located tens of kilometres south of the basin, from which the fans were scattered and covered only the southern part of the basin (Pimentel and Azevêdo 1995). These events have been considered to belong to the middle Vilafranquian (Azevêdo 1982a, b) and to be related to the Ibero-Manchega II tectonic phase (Calvo et al. 1993).

Quaternary tectonics subsequently reactivated major faults around the basin, creating high crests (above 200 m). However, the prevailing cold, damp climate did not promote the formation of alluvial fans, but rather the development of the modern drainage system, as indicated by several fluvial terraces (Pimentel and Azevêdo 1990).

8 The Algarve Cenozoic Platform and Guadalquivir Basin (Western Sector)

The first studies of the Cenozoic of the Algarve (Figs. 1, 2, 28 and 37) dealt only with a limited number of fossil localities and deposits (Costa 1866; Dollfus et al. 1903–1904; Bourcart and Zbyszewski 1940; Chavan 1940; Zbyszewski 1948; 1950). Some interpretational syntheses have since been published (Antunes et al. 1981, 1990b; Antunes and Pais 1992a, b). In addition, new developments have been made, based mainly on Sr isotopic dating at the Department of Earth Sciences at The University of Cambridge.

Table 2 Results of ^{87}Sr/^{86}Sr age determinations for the Neogene of Algarve (Pais et al. 2000)

Localities	Samp.	Ma	^{87}Sr/^{86}Sr	±2ρ
Sagres		12.0(+1.3,−1.3)	0.708886	0.000021
		12.5(+0.8,−0.7)	0.708850	0.000047
Alzejur Furna Amarela	3	9.8 ± 1.5	0.708903	0.000024
	2	12.6(+0.9,−0.8)	0.708850	0.000020
	1	11.9(± 0.8)	0.708867	0.000018
Alzejur Igreja Nova		16.9(± 0.3)	0.708701	0.000018
Alzejur Gimnosdesportivo	2	18.5(± 0.3)	0.708605	0.000017
	1	19.5(+0.3,−0.2)	0.708537	0.000017
Zavial		16.2(+0.5,−0.5)	0.708734	0.000021
Canavial beach	1	17.5(+0.7,−0.7)	0.708653	0.000026
	3	16.3(+0.5,−0.3)	0.708737	0.000020
	4	14.3(+0.5,−0.5)	0.708804	0.000017
	5	15.3(+0.5,−0.5)	0.708771	0.000018
Albardeira		5.5(+4.1,−0.7)	0.708959	0.000018
Rocha beach	2	12.2(+1.2,−1.3)	0.708860	0.000018
	3	11.5(+0.8,−0.5)	0.708875	0.000026
	II	10.7(+0.8,−1.2)	0.708886	0.000018
Albandeira	W7	15.5(± 0.4)	0.78763	0.000017
	W6	16.3(+0.4−0.5	0.708799	0.000017
Galé beach		11.3(+0.9,−1.3)	0.708880	0.000024
Arrifão	1	19.5(+0.2,−0.3)	0.708515	0.000017
	2	19.2(+0.2,−0.4)	0.708553	0.000018
	3	14.2(+0.6,−0.7)	0.708808	0.000026
	3	14.2(+0.6,−0.4)	0.708807	0.000020
	4	8.3(+2.2,−3.3)	0.708922	0.000021
Castelo beach		15.5(+0.8,−0.3)	0.708759	0.000017
Mem Moniz	Forams	12.5(−1.7,+0.7)	0.708857	0.000016
Santa Eulália beach	3	14.4(+0.5,−0.6)	0.708804	0.000017
	2	14.6(+0.5,−0.6)	0.708799	0.000017
	1	15.5(± 0.4)	0.708763	0.000017
Hotel Auramar beach	2	9.5(+1.0,−0.5)	0.708907	0.000016
Olhos de Água sands		3.0(+2.5,−1.0)	0.709044	0.000048
	Shark tooth	4.8(+0.7,−0.8)	0.708797	0.000020
Quelfes	4	5.2(+4.4,−0.6)	0.708961	0.000014
	3	5.2(+4.4,−1.1)	0.708971	0.000018
	1	5.3(+4.3,−0.7)	0.708952	0.000018
Cacela	1	5.7(+3.9,−1.1)	0.708950	0.000014
	4	8.3(+2.2,−3.3)	0.707922	0.000021

Current knowledge concerning the Paleogene in the Algarve appears to be limited to Guia conglomerates and mudstones (Fig. 28). The Neogene units characterized thus far are mostly of Miocene age and mostly marine. They developed mainly in a temperate carbonate platform environment until the end of the Middle Miocene. Thereafter, the region became dependent on the Guadalquivir

Basin, which constitutes the western border of the area, mainly to the east of the Quarteira fault. The Pliocene, according to $^{87}Sr/^{86}Sr$ age determinations and faunal assemblages, seems to be represented by the Falésia sands (Ludo Formation), and those of Morgadinho (Antunes et al. 1986a) are, at least in part, of the same age. The Quarteira sands (also known as the Faro-Quarteira Formation) mainly correspond in age to the Early Pleistocene. Very recent sandy deposits, such as at Guia, host a fauna of small vertebrates (Antunes et al. 1989). Lithic artefacts within those deposits corroborate a probable latest Pliocene—Early Pleistocene age (Cardoso et al. 1985). Ages of other Quaternary deposits can be inferred according to their stratigraphic position, such as karst deposits and fluvial and marine terraces. Regarding the terraces, there is an important deposit containing mammals in Algoz (ante-Günz glaciation, Bihariano) (Antunes et al. 1986a) and also Mousterian lithic industries. Two karst infills near Goldra (Riss-Würm interglaciation or one of the first Würm interstades; Antunes et al. 1986b) have provided artefacts and vertebrates (Algueirão da Goldra, Holocene, Póvoas et al. 1995).

Establishing stratigraphic relationships between the sedimentary units of the Neogene in geographically remote areas is difficult due to the occurrence of erosional episodes and varied sedimentary facies. However, the effort placed on dating (Table 2) has contributed substantially to understanding the evolution of the Algarve Basin during the Neogene.

Although Sousa (1917, 1922) reported the presence of Miocene igneous rocks in the Algarve, this has not subsequently been verified. In 1971, a small outcrop of Miocene basanite embedded in biocalcarenite was studied at Figueira (Portimão), and showed evidence of low-grade thermal effects along the contact with the igneous dike. Those igneous rocks may represent, therefore, the most recent manifestation of volcanic activity in Portugal (Coelho and Bravo 1983). Cenozoic volcanic activity has been established for the Langhian-Serravallian to the Quaternary in southern Spain.

The tectonic structures of the Mesozoic and Cenozoic deposits in the Algarve allow the following tectonic episodes of the Alpine orogeny to be recognized: (i) Jurassic (late Triassic at least) to Early Cretaceous N–S extension; (ii) E–W distension during the Early to Late Cretaceous; (iii) N–S compression during the evolution of the Monchique Syenite massif in the Late Cretaceous; (iv) Paleogene compression (Albufeira saline dome); (v) Strain N–S and E–W in the Early and Middle Miocene; (vi) Tortonian compression N–S and E–W compression post-Messinian; (vii) N–S compression during the Quaternary (Kullberg et al. 1992).

NE–SW fractures affecting the Paleozoic basement are related to the first extensional stages whereas the N–S Mesozoic distension is mainly responsible for the two E-W flexures occurring in the Algarve Basin. The transition to a compressive tectonic regime occurred after the Monchique massif intrusion. If the angular unconformity between the Lower Cretaceous and Lower Miocene in Arrifão (Albufeira) is due to a local halokenesis effect, then the real Algarve Basin tectonic inversion occurred only in the early to middle Tortonian (Kullberg et al. 1992; Terrinha 1998; Dias 2001; Lopes and Cunha 2007; Lopes et al. 2008a, b).

8.1 Lower and Middle Miocene

Carbonate sedimentation prevailed during the Early and Middle Miocene, becoming essentially fine-grained and siliciclastic in the Late Miocene, and showing strong facies affinities with the sedimentary facies infilling the Guadalquivir basin at Spain (Pais et al. 2000; Legoinha 2001).

The *Lagos-Portimão Formation* (LPF) is ∼ 60 m thick and consists of sandstones and carbonate rudstones deposited in a mixed carbonate-siliciclastic platform during the Early and Middle Miocene (Antunes et al. 1990b; Antunes and Pais 1992a, b; Pais et al. 2000; Legoinha 2001; Forst 2003; Brachert et al. 2003; Kroeger et al. 2007) (Figs. 28, 29, 30, 31 and 37, Table 2). The deposits have been the subject of considerable paleo-environmental interest. Previous studies (Forst 2003; Brachert et al. 2003; Pais et al. 2000) considered the LPF to have been accumulated along a narrow shelf platform and influenced by frequent eustatic cycles leading to synchronous facies variations over the platform. The LPF's main lithotype is a very fossiliferous, massive yellow or pink biocalcarenite. At the base, the fossil assemblage is dominated by molluscs, whereas further up the unit, species include sea urchins, bryozoans, pectinidae and fish. A layer rich in *Celleporaria palmata* (bryozoan) in biocenose with the coral *Culizia parasitica* is well exposed at the top of the coastal cliffs in the Castelo and Coelha sectors. Additionally, some other layers with corals occur within the LPF. Both the reported faunal assemblage and a noteworthy metric layer of rodolithic facies point to warm temperate climatic conditions in a shallow marine environment.

The LPF deposits are arranged geometrically in rather continuous layers, and are well exposed in the coastal cliffs of the Algarve, from Lagos to Olhos de Água; they have had considerable paleo-environmental interest. They overlap the Carboniferous, Jurassic, Cretaceous and possible Paleogene units. In some localities, the boundary between the LPF and the older geological units is through angular unconformity, while in other places hiatus and/or paraconformities have been recognized. The LPF is disconformably overlain by Tortonian marine siliciclastic sediments.

The tabular character of the LPF, which allows a remarkable exposure for 40 km along the coastal cliffs of the Algarve region, as well as the unit's thickness and lithological monotony, are the result of the sedimentary facies being deposited essentially parallel to the direction of stretching of the platform (Brachert et al. 2003; Kroeger et al. 2007). The exposed sections are considered representative of the entire unit. However, sedimentological and stratigraphic details suggest a more complex genesis.

Four reference marker beds (mb) (Forst et al. 2000; Forst 2003), easily recognizable in the field, have been characterized:

(i)   The lowest marker bed I (mb I, not reported by Kroeger et al. 2007) is a red or brown biocalcarenite layer, very rich in mollusc fragments (pectinidae and oysters). Due to its basal position, it rarely crops out above sea level, but is observable in the region of Carvoeiro.

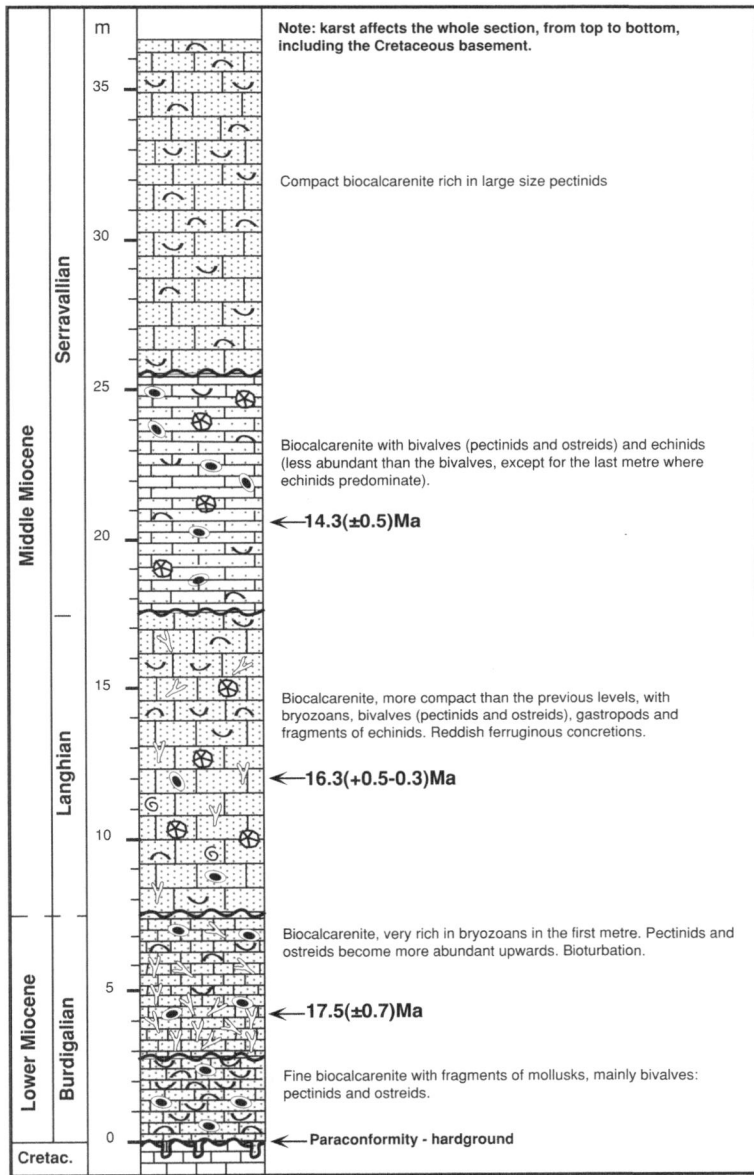

Note: karst affects the whole section, from top to bottom, including the Cretaceous basement.

Compact biocalcarenite rich in large size pectinids

Biocalcarenite with bivalves (pectinids and ostreids) and echinids (less abundant than the bivalves, except for the last metre where echinids predominate).

←—14.3(±0.5)Ma

Biocalcarenite, more compact than the previous levels, with bryozoans, bivalves (pectinids and ostreids), gastropods and fragments of echinids. Reddish ferruginous concretions.

←—16.3(+0.5-0.3)Ma

Biocalcarenite, very rich in bryozoans in the first metre. Pectinids and ostreids become more abundant upwards. Bioturbation.

←—17.5(±0.7)Ma

Fine biocalcarenite with fragments of mollusks, mainly bivalves: pectinids and ostreids.

←— Paraconformity - hardground

Fig. 29 Canavial beach section (Pais et al. 2000)

(ii)   Marker bed II (mb I of Kroeger et al. 2007) is a yellowish to brownish sandstone, particularly rich in bryozoan *Calpensia* and branched colonies of celeporids.

(iii) Marker bed III (mb II of Kroeger et al. 2007) is also a yellow fossiliferous
       sandstone whose base is strongly bioturbated and is rich in foliated bry-
       ozoans and macroforaminífera. Masses of Heterostegins 2–3 cm in diameter
       are also present in this marker.

(iv)   Marker bed IV (mb III of Kroeger et al. 2007) is composed of three layers: a
       fossiliferous sandstone intercalated with two sandstones. The upper sand-
       stone is rich in bryozoan *Celleporaria palmata* associated with the coral
       Culizia parasitica.

$^{87}Sr/^{86}Sr$ isotope determinations of the described marker beds have constrained the age of the portion of the LPF in which they occur to the Burdigalian
(Table 3).

Other $^{87}Sr/^{86}Sr$ age determinations obtained for different sections of the LPF
have identified the existence of hiatuses between 15.6 and 16.3 Ma and between
13.1 and 14.4 Ma (Kroeger et al. 2007), corresponding respectively to the
Langhian and early Serravallian (Table 2). The only known exception is the dense
cluster of disarticulated oysters from Arrifão, dated at 14.2 (+0.6,−0.7) Ma, which
may correspond to a local event. These gaps are contemporaneous with Division
Vb from Lisboa, with part of the hiatus being recorded in the lower part of Penedo
North section and with the episode prior to the deposition of the glauconitic layer
from the same section. Therefore, this hiatus will be a little younger than the
Arrábida tectonic episode (Forst et al. 2000; Forst 2003), easily recognizable in the
field. The upper Serravalian sedimentary record is preserved according to the
$^{87}Sr/^{86}Sr$ isotopic ages of 12.2 (+1.2,−1.3) Ma and 11.5 Ma (+0.8,−0.5) Ma
(Table 2), being exposed in the Rocha Beach (R) coastal cliff (Fig. 31) where
Legoinha (2001) recognized the follow foraminifera:

R1—*Orbulina universa, Orbulina suturalis, Globoquadrina*;
R3—*Orbulina, Globigerinoides triloba, Globigerina praebulloides, Globigerina
angustiumbilicata*;
R4—*Orbulina, Globigerinoides triloba, Globigerina angustiumbilicata*.

In addition, the presence of *Orbulina universa* in a borehole drilled in Faro
(Central Algarve) and at a section of Vau Beach, clearly indicates that sedimentation in the Serravallian is represented (Antunes et al. 1984; Antunes and Pais
1992a, b).

The reported $^{87}Sr/^{86}Sr$ ages (Tables 2 and 3) corroborate early studies carried
out by Antunes et al. (1981) which recognized foraminiferal fauna, among them
Globigerinoides triloba, Globigerinoides subquadratus, Praeorbulina (doubtful),
Globigerinoides bisphericus (?) and *Praeorbulina transitoria* pointing to late
Burdigalian (N7) or possibly early Langhian (N8) ages for the lower part of
the LPF.

Legoinha (2001) recognized the following benthic foraminiferal assemblages
characteristic of infralittoral environments:

R1—*Cibicides, Guttulina, Heterolepa, Nonion, Textularia*.
R3—*Ammonia, Globulina, Guttulina, Lenticulina, Nonion, Quinqueloculina*.

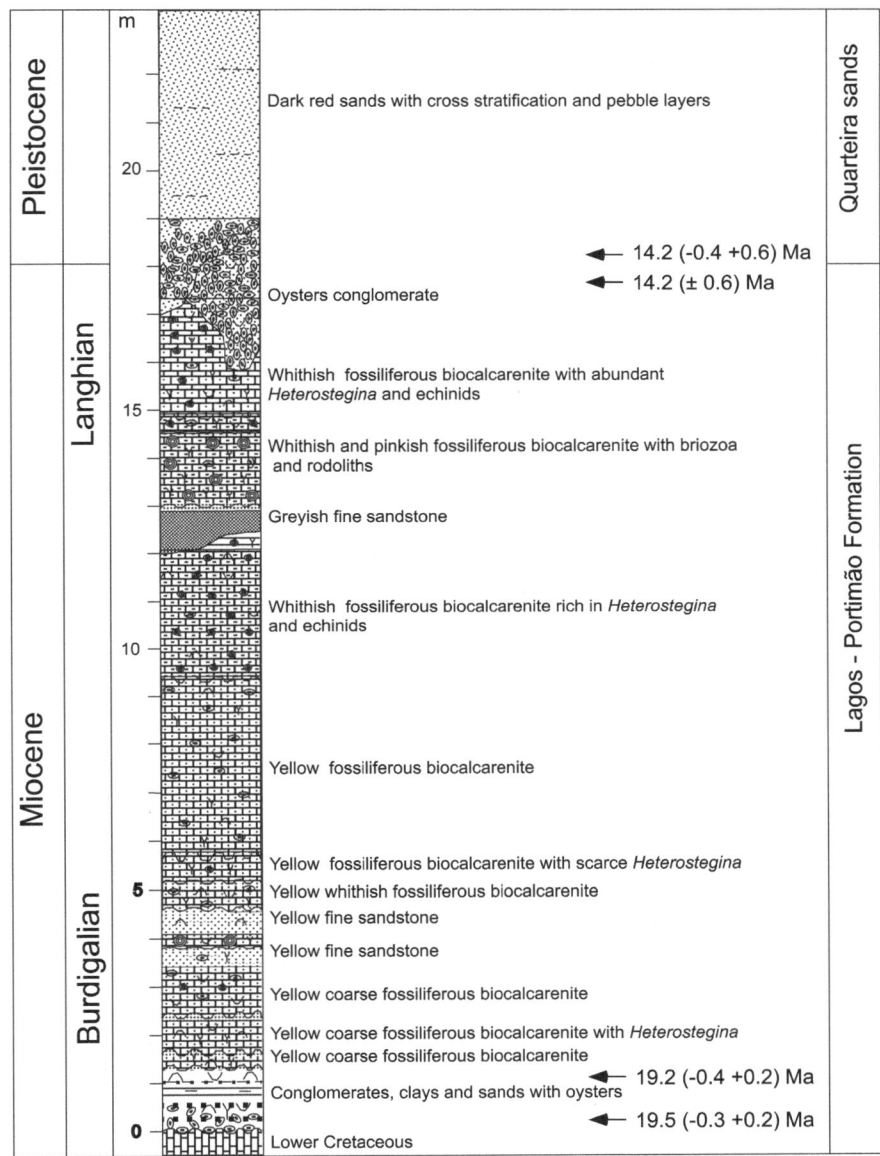

Fig. 30 The Arrifão stratigraphic section (updated from Pais 1982; Legoinha 2001)

The Lagos-Portimão Formation exposed in the Algarve coastal cliffs was deposited between the early Burdigalian and the late Serravalian according to the $^{87}Sr/^{86}Sr$ isotopic age determinations (Table 2). The uppermost part of the LPF is well represented at Galé (midway between Albufeira and Armação de Pêra) where

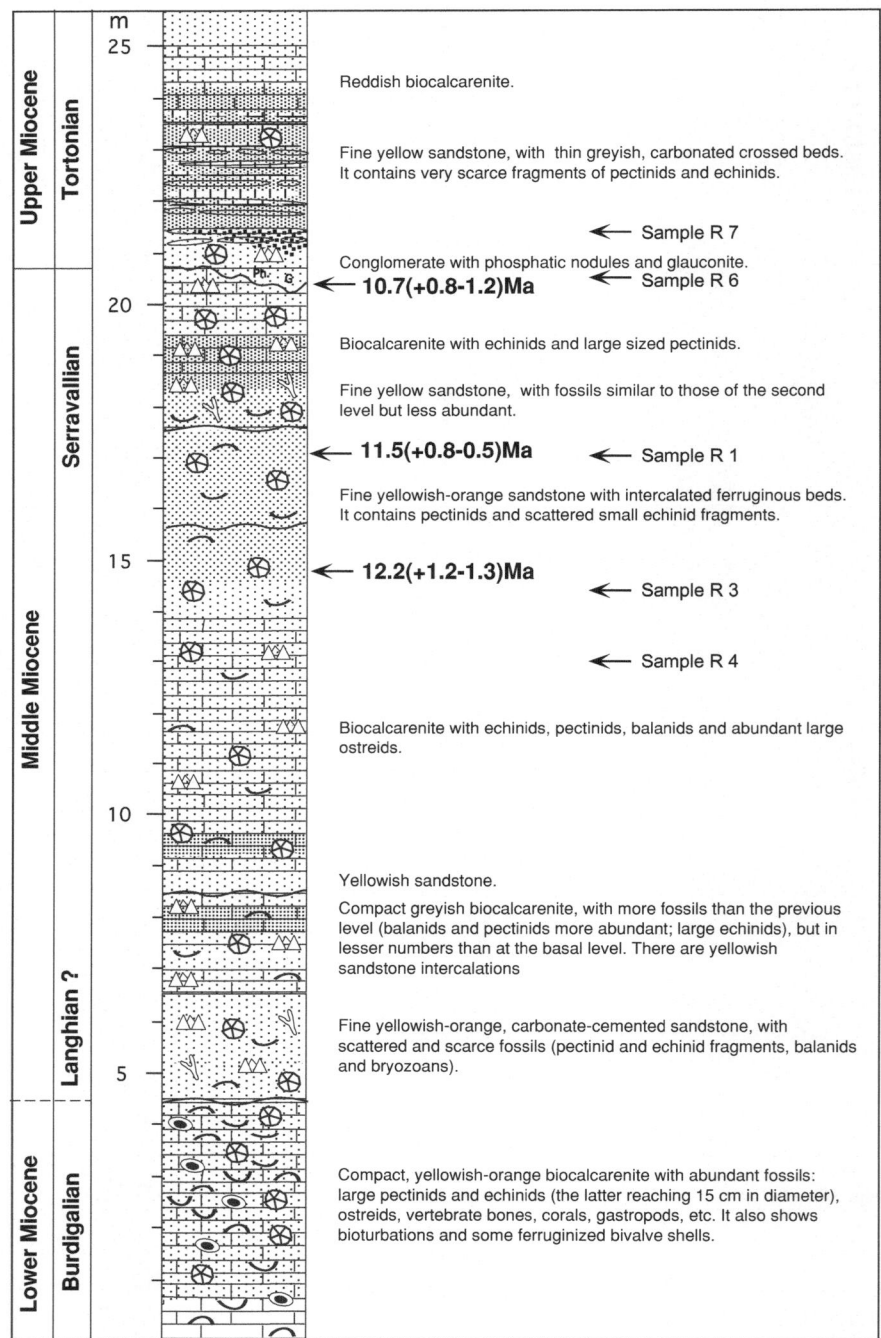

Fig. 31 The Rocha beach stratigraphic section (Pais et al. 2000; Legoinha 2003)

Table 3 ^{87}Sr/^{86}Sr dating of the Lagos Portimão Formation marker beds

Marker bed	^{87}Sr/^{86}Sr dating (Ma)	Author
mb I	17.75 (+0.20,−0.19)	Breisig 2000 *in* Forst 2003
mb II	16.70 (+0.25,−0.29)	Breisig 2000 *in* Forst 2003
mb III	16.56 (+0.26,−0.31)	Breisig 2000 *in* Forst 2003
mb IV	15.94 (+0.38,−0.41)	Breisig 2000 *in* Forst 2003
	15.5 (+0.8,−0.3)	Pais et al. (2000)

a layer of biocalcarenites ∼ 8 m in thickness belongs to the upper part of the formation. At the base, there is a 3.4-m-thick layer of limestone rich in rhodoliths. Towards the top, sand and shells of molluscs (Pectinidae, *Spondylus*) both increase in frequency. There are also colonies of bryozoa nodular celeporiformes, large foraminifera (*Heterostegina*) and sea urchins (*Clypeaster*). The upper sandstones contain some Pectinidae, balanids in small colonies and fragments of *Pinna*. The set culminates with a grey sandstone layer, 40 cm thick, bearing quartz pebbles and overlying the biocalcarenite through an irregular erosive surface. Pectinidae, oysters, large *Balanus* and sea urchins are abundant.

The surface of the biocalcarenites is deeply karsified and contains several cavities filled by red mudstones as detrital residues after the carbonate dissolution. The paleorelief inherited from an erosional phase affecting the LPF is filled by fine sandstones with glauconite, where K–Ar ages suggest early Tortonian (10.1 ± 0.25 Ma; Antunes et al. 1984) or late Tortonian (8.15 ± 0.29 to 7.54 ± 0.27 Ma; Boski et al. 1995).

A lenticular body of considerable lateral dimension and composed of white, light sands displaying cross-bedding and numerous other sedimentary structures occurs at Olhos de Água. Westwards of this locality, at Maria Luisa, the thickness of the sandy unit decreases progressively, and biocalcarenites, the most common lithological type of the Lagos-Portimão Formation, are developed. ^{87}Sr/^{86}Sr determinations of mollusc shells from layers below the sandstones have given an age of 15.5 ± 0.4 Ma, while the shells collected in the sandstones have given 14.6 +0.5,−0.6 Ma, and the layers above have given 14.4 +0.5,−0.6 Ma.

Nanoplankton (*Reticulofenestra pseudoumbilica, Calcidiscus premacintyrei*) from the top of the LPF (Albardeira, Lagos), also points to the Serravallian (NN6 or CN5a) (Cachão, 1995). Moreover, the unit is older than the glauconitic sands of Galé beach, which provided K–Ar ages of 10.1 ± 0.25 Ma, and than the sandstones of Rocha beach, dated using ^{87}Sr/^{86}Sr isotopes at 10.7 +0.8, −1.2 Ma.

Biocalcarenites exposed in the Rocha beach cliffs are fairly consolidated, corresponding to a coastal environment of high energy according to the biosedimentary facies (Fig. 30). They have been attributed to the "Helvetian" based on molluscs (Cotter *in* Dollfus et al. 1903–1904) and are correlated with Divisions Vb and VIc from the Lisboa region (Ferreira 1951). The unit culminates with an erosional surface and is overlain by laminated sandstones, with few fossils. The first layer of these laminated sandstones is a conglomerate containing fragments of phosphate and glauconite crusts. Both Antunes et al. (1997b) and Pais et al. (2000) described the section and reported isotopic ages (Fig. 30).

Field observations between Vau and Rocha beaches have shown that while the thickness of these facies remains constant, the texture and fossil content change significantly. The maximum abundance of fossils varies between 55 and 75%, with concentrations of clypeasteroid echinids, bryozoans, oysters, Pectinidae and gastropod casts. This is particularly evident in the transition from coarse rudstones with skeletal remains (SR in the terminology of Bachert et al. 2003) that constitute the top and middle of the section at Rocha beach to fossiliferous sandstone (FS) and SR sequences arranged in coarsening-upward sequences in Vau beach (Dabrio et al. 2008).

In general, the microfacies are thin and include sandstone interspersed with a few SR levels. Some samples are concentrated in poorly-sorted fossil remains. The following components are common in the deposits: quartz (silt to medium sand reaches 5–25%), authigenic and detrital glauconite, pellets, foraminifera (*Globigerina, Globigerinoides, Heterostegina*, Miliolids, Nonionids, Textulariidae), echinoids, molluscs (oysters, Pectinidae) and some gastropods, bryozoans, (more abundant in thin facies, with fewer skeletal components), brachiopods, ostracods and green algae. The cement includes a mosaic of anhedral inequigranular calcite.

Between Rocha and Vau beaches, it has proved possible to distinguish five subunits (Dabrio et al. 2008):

Subunit 1 Fossiliferous sandstones and coarse sandstones with large-scale crossbedding. *Thalassinoides* are plentiful and blur the original structures. Interstratal karsification in some layers also complicates the structure. The thickness of the cross-bedded layers exceeds 1 m, and the arrangement and geometry of sedimentary bodies indicates a migration of bedforms fuelled by fossiliferous sand accumulated in shallow water off the coast. The absence of lutites is attributed to the entrainment of fines by currents. The overall pattern indicates bars in a high-energy coastal shelf environment subject to regional west-flowing currents.

Subunit 2 Includes two intervals: the lower (2.5 m thick) consists of a yellow sandstone containing *Thalassinoides*, other fossils are scarce. At Rocha beach, skeletal particles are larger than to the west (W Rocha and Vau beaches) and the interval consists of cross-bedded skeletal rudstones (SR). The upper part is variable in composition: Rocha beach is of SR type, but the particle size decreases towards Vau beach; at Rocha beach this changes to vertical stacking fossiliferous sandstones sequences (FS) and SR with increasing particle size to the top, with colors ranging vertically from whitish to grey and reddish. The top is hardened, deeply punctured, and SR facies has been considered an hardground. The general pattern consists of a shallow marine accumulation of intensely-burrowed skeletal packstone and grainstone that changes to finer-grained, burrowed, coarsening-upwards sequences with blurred bedding. This is interpreted to reflect the transition, in a SW direction, to deeper settings on a ramp where currents scarcely swept the bottom and wave winnowing was much reduced.

Subunits 3, 4 *and* 5 These consist of thick stacks with metric-scale layers with FS facies and FR changing laterally to a repetition of decimeter-scale thickening-upwards sequences of type FS and SS as observed in subunit 2.

It is possible to recognize three types of sequences. At the small scale, there are layers representing storm events, locally preserved although bioturbation. A second order of sequences observed in subunits 3 and 5 at Vau beach are dominated by yellow sand at the bottom (FS) with bryozoans and *Thalassinoides*. Locally, wavy lamination is almost completely obliterated. They transition gradually to SR facies with large rhodoliths, *Pecten, Ostrea*, echinids (*Scutella, Clypeaster*) and *Balanus*. These sequences are considered to have resulted from Milankovitch cycles. The bottom of the cycle corresponds to wetter conditions and more storms with a large supply of sand. The upper part represents the most arid intervals, with a reduced terrigenous supply favouring the accumulation of shells (Dabrio et al. 2008).

The larger-scale sequences are represented by stacks of repeating sub-units corresponding to rises in sea level at the bases of yellow FS layers. These are marked by the accumulation of second-order sequences, apparently representing intervals of time with greater stability of sea level. The three-dimensional architecture and internal organization of the sequences are thought to record rising sea levels but separated by intervals of greater stability in sea level with a strong component of orbital forcing, although this has not yet been conclusively proved. Moreover, local variations in bathymetry favoured the occurrence of shallower areas where currents and waves disrupted coarser clastic facies, while in the deeper areas, energy of tempests with Milankovitch forcing are better recorded (Dabrio et al. 2008).

8.1.1 Paleogeography

The Miocene Mediterranean Sea was linked to the Atlantic Ocean through the narrow North Betic and South Riffean straits. It was linked to the Indo-Pacific ocean to the east until the collision of the African and European plates closed the crossing, making the Mediterranean a marginal sea, and changed the pattern of circulation in the North Betic strait. A model for the Middle Miocene indicates estuarine facies circulation where diatomite (moronites) testify to the abundance of nutrients and high productivity in the southern flank of the North Betic strait (Sierro et al. 1989; Sierro and Flores 1992). The model assumes a Mediterreanean outflow system of surface water circulation to the Atlantic with a general SW direction and a deep flow back from the Atlantic to the Mediterranean. The orientations of current directions measured in subunits 1 and 2, and the general model of dispersion systems, agree with the standard estuarine pattern (Dabrio et al. 2008).

Kroeger et al. (2007) made a comparison of the biocalcarenites of Lagos-Portimão with equivalent rocks from Crete. They believe that the wide distribution of the sandstone-limestone sets, sedimentary lithological marker beds and surfaces suggests that basins of the Central Algarve and Crete were influenced by changes in relative sea level that affected the entire region with a strong eustatic component.

A key point for the proposed model is to explain the changes in bathymetry responsible for the deposition of the Neogene units and generation of discontinuities on the Algarve platform; these bathymetric factors include variations in sea level on a large scale, and local irregularities such as salt domes between Portimão and Albufeira. These latter irregularities created a slight bend which restricted the depth and therefore induced local changes in facies, even over short distances, but no significant variations in thickness, according to the mechanism suggested by Forst (2003). In addition, Dabrio et al. (2008) proposed that the currents that swept the platform tended to accelerate in headlands or in shallow areas, eroding the bottom and depositing carpets of particles where the flow and transport capacity slowed down, as demonstrated by Swift and Thorne (1991). In southern Portugal, the modern streams flowing out of the North Betic strait transport sediments to the west, extending the Coriolis effect during storms. The direction of change of facies described for the LPF is to the south–west.

In general, the LPF is well exposed between the Olhos de Água and Canavial beaches along the coastline, and crops outs inland until north of Bensafrim. At the Aljezur graben, there are also some detrital deposits and biocalcarenites of the same age as the LPF. The deposition seems to have been controlled by the S. Marcos-Quarteira fault.

8.2 Upper Miocene

Upper Miocene deposits are particularly well represented in the eastern Algarve (Figs. 28, 32, 33 and 37), but also crop out near Lagos (Meia Praia) in the western Algarve. They include, primarily in the eastern Algarve, sandstones and fine sand with some glauconite. In the western Algarve, the Upper Miocene is represented by limestones with quartz pebbles, which may be equivalent to the base of the formation cropping out at both Cacela and Quelfes, and has also been also recognized in the region of Meia Praia, near Lagos.

In the central Algarve, in Mem Moniz, spongoliths crop out, although their age is not particularly well constrained. The calcareous nannofossils and Sr isotopes indicate these deposits are upper Serravallian, while planktonic foraminifera indicate Tortonian.

Sandstones and fine sands of Rocha and Auramar Hotel beaches disconformably overlie the LPF (Antunes et al. 1981; Pais et al. 2000). At Rocha beach, the base consists of conglomerates with phosphate clasts and glauconite. At Auramar Hotel beach, fine yellowish sands are exposed.

8.2.1 Fine Sands and Sandstones

This unit is separated from the underlying layers of the LPF by a regional disconformity. It corresponds to a significant shift from carbonate to siliciclastic sedimentation. This modification can be matched to the beginning of the second-order eustatic cycle T3 (Haq et al. 1987). The lowest layer at Rocha beach is a conglomerate with phosphate clasts.

At Rocha beach (R), layers above the surface of discontinuity that separates fine sandstones from the LPF have provided the following planktonic foraminifera (Legoinha 2001):

R6—*Dentoglobigerina, Globigerinoides triloba, Globigerina bulloides, Globigerina angustiumbilicata*;
R7—*Orbulina, Globigerina bulloides, Globigerina concina, Globigerinella aequilateralis, Globigerinoides triloba*, cf. *Neogloboquadrina acostaensis* (five specimens).

The presence of *Globigerina concina* and *Neogloboquadrina acostaensis* point to the Upper Miocene, Tortonian (N16) or later. Also, the dating of oyster shells from the basal layer of conglomeratic sandstones yields a $^{87}Sr/^{86}Sr$ age of 10.7 +0.8, −1.2 Ma (early Tortonian).

Benthic foraminifera are more abundant than in the LPF, and include (Legoinha 2001):

R6—*Globulina spinosa* (abundant); *Cancris, Lagena, Hanzawaia* and *Heterolepa* (all abundant); *Ammonia, Gyroidina, Neoconorbina, Nodosaria, Pullenia, Uvigerina, Textularia* (scarce).
R7—*Globulina, Textularia* (abundant*); Bulimina, Cancris, Hanzawaia, Nodosaria, Nonion* (all abundant); *Lagena* (scarce).

These suggest environments a little deeper than those of samples R1 and R3 (Legoinha 2001).

Fine sands from Auramar Hotel beach, dated at 9.5 +1.0, −0.5 Ma, are interpreted as correlative of this unit. At Galé beach, fine sand rich in glauconite yielded a K–Ar age of 10.1 (± 0.25) Ma (early Tortonian) (Antunes et al. 1984). Boski et al. (1995) obtained for the same deposits in the region of Galé values of 8.15 ± 0.29 and 7.54 ± 0.27 Ma.

In Campina de Faro and Olhão, there are outcrops of fossiliferous limestones, rich in rounded quartz pebbles; in some places, these deposits change gradually to micro-conglomerates (Faro limestone with quartz pebbles and conglomerates). At the top of Auramar Hotel Beach, there is an initial layer of conglomerates with oysters that have provided a $^{87}Sr/^{86}Sr$ age of 8.3 +2.2, −3.3 Ma, concordant with the biostratigraphical dating of the basal conglomerate cropping out in the Cacela creek (biozone N16-N17).

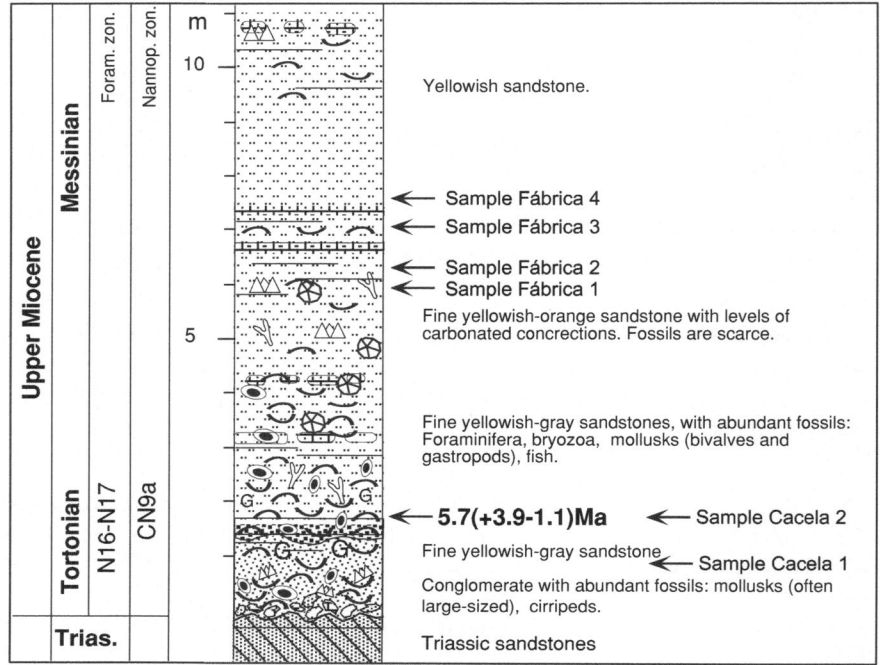

Fig. 32 The Cacela stratigraphic section (Pais et al. 2000; Legoinha 2003)

8.2.2 Cacela Formation

The Cacela Formation (Fig. 32) consists of conglomerates at the bottom, fine sand and carbonate-rich sandstone, with molluscs, fish, ostracods, foraminifera, palynomorphs and calcareous nanoplankton. In Cacela creek, the lower layers (fine sands and conglomerates) can be observed overlying the Triassic by angular unconformity. At Fábrica, to the south-west, the upper layers (yellow silts) can be observed, representing the Lower and Middle Members of the Cacela Formation (Antunes et al. 1981).

In these deposits are contained the richest Portuguese site of Miocene molluscs, in diversity, quality and abundance. They have been studied in numerous investigations including Costa (1866), Cotter (1879; in Dollfus et al. 1903–1904; in Choffat 1950), Chavan (1940), Bourcart and Zbyszewski (1940), Freneix (1957), Brébion (1957), Santos and Boski (1998), Santos et al. (1998) and Santos (2005).

Antunes et al. (1981) reported the presence of *Globigerinoides extremus*, *Globorotalia* aff. *conomiozea*, *Globorotalia pseudomiocenica*, *Globotalia menardii*, *Globorotalia acostaensis* (sinist.) e *Globorotalia humerosa* (sinist.) indicating the top of zone N16 or, more likely, N17.

At the Fábrica outcrop, Antunes et al. (1990b) identified abundant planktonic foraminifera, including *Globigerinoides extremus, Globigerinoides seigliei* and

Neogloboquadrina acostaensis (sinist.). Considering the presence of benthic foraminifera *Spiroplectammina carinata*, which disappears in the basal Messinian, the deposits can be assigned to the upper Tortonian. In addition to this species, Antunes et al. (1981) reported *Marginulina, Heterolepa, Ammonia, Uvigerina* and *Nonion* in the Cacela creek outcrop.

Legoinha (2001, 2003) first recognized the occurrence in Cacela creek 2 of *Globorotalias* of the *menardii* group with sinistral winding. Sierro (1985) and Sierro et al. (1993) had earlier identified a series of bio-events in the Guadalquivir Basin. The first is defined by the sudden disappearance or reduction of *Globorotalia menardii* group I (sinistral). The second is marked by the appearance of abundant *Globorotalia menardii* group II (dextral). Between these two events, the temperate waters of the North Atlantic and the Mediterranean were practically devoid of keeled globorotalias (Sierro et al. 1993).

Fábrica 2 provides a rich and diverse association of foraminifera: *Globigerina bulloides, Globigerina apertura, Globigerina drury, Globigerinita glutinata, Globigerinoides bulloideus, Globigerinoides extremus, Globigerinoides seigliei, Globigerinoides immaturus, Orbulina universa, Orbulina suturalis, Globigerina quinqueloba, Globoquadrina globosa, Globorotalia scitula, Neogloboquadrina acostaensis* (sin.). A lack of keeled globorotalias is evident (Antunes et al. 1990b).

Concerning benthic foraminifera, the following were recognized by Legoinha (2001):

Cacela creek 2: *Ammonia, Nonion, Elphidium, Cancris, Reussella, Nodosaria* (abuntant); *Guttulina, Lenticulina, Fursenkoina* and *Neoeponides schreibersi* (rare); Fábrica 2: *Ammonia, Nonion, Textularia, Spiroplectaminna carinata, Brizalina, Bulimina, Bolivina, Cibicides, Globobulimina, Lenticulina, Pullenia, Guttulina e Marginulina.*

Antunes et al. (1981) indicated the presence of *Spiroplectaminna carinata, Marginulina, Heterolepa, Ammonia, Uvigerina* and *Nonion* in the outcrops of Cacela creek. Investigation of microfauna in the intermediate (yellow silts) and upper (grey silts) layers from the Cacela creek section did not provide any noteworthy results.

Civis et al. (2000) indicate that in the layers with concentrations of molluscs the benthic foraminifera are abundant but less diverse. *Ammonia beccarii* and *Nonion boueanum* predominate. *Elphidium crispum* and *Lobatula lobatula* occur to a lesser extent, with some Discorbidae and Nodosaridae (large shells). The Fábrica benthic association is the most abundant and diverse (some hundreds of species), highlighting buliminides, bolivinides, uvigerinides and *Valvulineria bradyana.* These benthic foraminifera suggest shallow marine environments, with increasing depth for the section at Fábrica, and the development of confined environments with anoxic conditions.

Nanoplankton of the lower layers include: *Discoaster berggrenii, Helicosphaera stalis, Minylitha convalis, Triquetrorhabdulus rugosus* (?), suggesting assignation to the upper Tortonian (CN 9a zone of Bukry) (Cachão 1995). Slightly higher layers at the Fábrica section also contain calcareous nannofossils

(*Coccolithus pelagicus, Reticulofenestra pseudoumbilica, R. minuta, R. minutula, R. haqii, Dictyococcites antarcticus, Sphenolithus abies, S. moriformis, Helicosphaera carteri, Eudiscoaster surculus, E. icarus, E. intercalaris, E. pseudovariabilis, Triquetrorhabdulus rugosus*) and planktonic foraminifera (*Globigerina bulloideus, G. apertura, G. druryi, G. quinqueloba, Globigerinoides extremus, G. seigliei, G. quadrilobatus, Globigerinita glutinata, Orbulina universa, O. suturalis, Globoquadrina globosa, Globorotalia (Hirsutella) scitula,* and *Neogloboquadrina acostaensis* sin.) that can be attributed to the late Tortonian-Messinian (Antunes et al. 1990b).

The exposures at Cacela and Fábrica can be correlated with deposits from the Guadalquivir Basin located between events 1 and 2 of Sierro (1993), dating from the late Tortonian, and now dated at 7.51 and 7.35 Ma, respectively (F. Sierro, Univ. Salamanca, personal communication). Antunes et al. (1981) considered that the rich fauna of ostracods (with *Aurila (Cymbaurila) diecci, A.* gr. *semilunata, Carinocythereis galilea, Nonurocythereis seminulum*) suggests a Messinian age.

The intermediate layers of Cacela Formation are silty and rich in glauconite, and dated from the late Tortonian at 6.90 ± 0.18 Ma (Galvana), 6.88 ± 0.5 Ma (Quelfes) and 7.3 ± 0.4 Ma (near Luz de Tavira) (Antunes et al. 1984, 1986c, 1990b, 1992b). Higher layers are almost devoid of fossils.

Further west, at Olhão, Quelfes (Fig. 33) and near Faro, very bioturbated fine sands and marly silts occur. At Quelfes, a glauconitic layer crops out. Poorly-carbonated cement-rich conglomerates with siliceous pebbles and blocks of limestone are also exposed. Foraminifera are common to abundant, and diversity is high. The highest abundance and diversity of foraminifera, sample Q4, indicates a maximum depth (Legoinha 2001, 2003). *Neogloboquadrina acostaensis* (predominantly sinistr.), *Globigerinoides extremus* and some forms of keeled *Globorotalia* (predominantly dext.) indicate the final Tortonian (N17). Legoinha (2001) notes, however, that higher layers at Quelfes, with *G. miotumida* and *G. conomiozea*, indicate the Messinian (N17). In comparison with the Guadalquivir Basin, these deposits were positioned between events 2 and 3 (Sierro 1985; Sierro et al. 1993, 1996). Associations are characteristic of an infralittoral marine environment, with significant depth.

At Quelfes (Q) (Fig. 33), samples Q2 and Q4 are the richest in foraminifera. In sample Q3, planktonic foraminifera are uncommon.

Species have been determined as follows (Legoinha 2001, 2003):

> Q1—*Dentoglobigerina altispira, Globigerinoides bulloideus, Globigerinoides conglobatus, Globigerinoides extremus, Neogloboquadrina humerosa, Orbulina universa.*

> Q2—*Dentoglobigerina altispira, Globigerina bulloides, Globigerina praecalida, Globigerinoides bulloideus, Globigerinoides conglobatus, Globigerinoides extremus, Globigerinoides seigliei, Neogloboquadrina humerosa, Orbulina suturalis, Orbulina universa.*

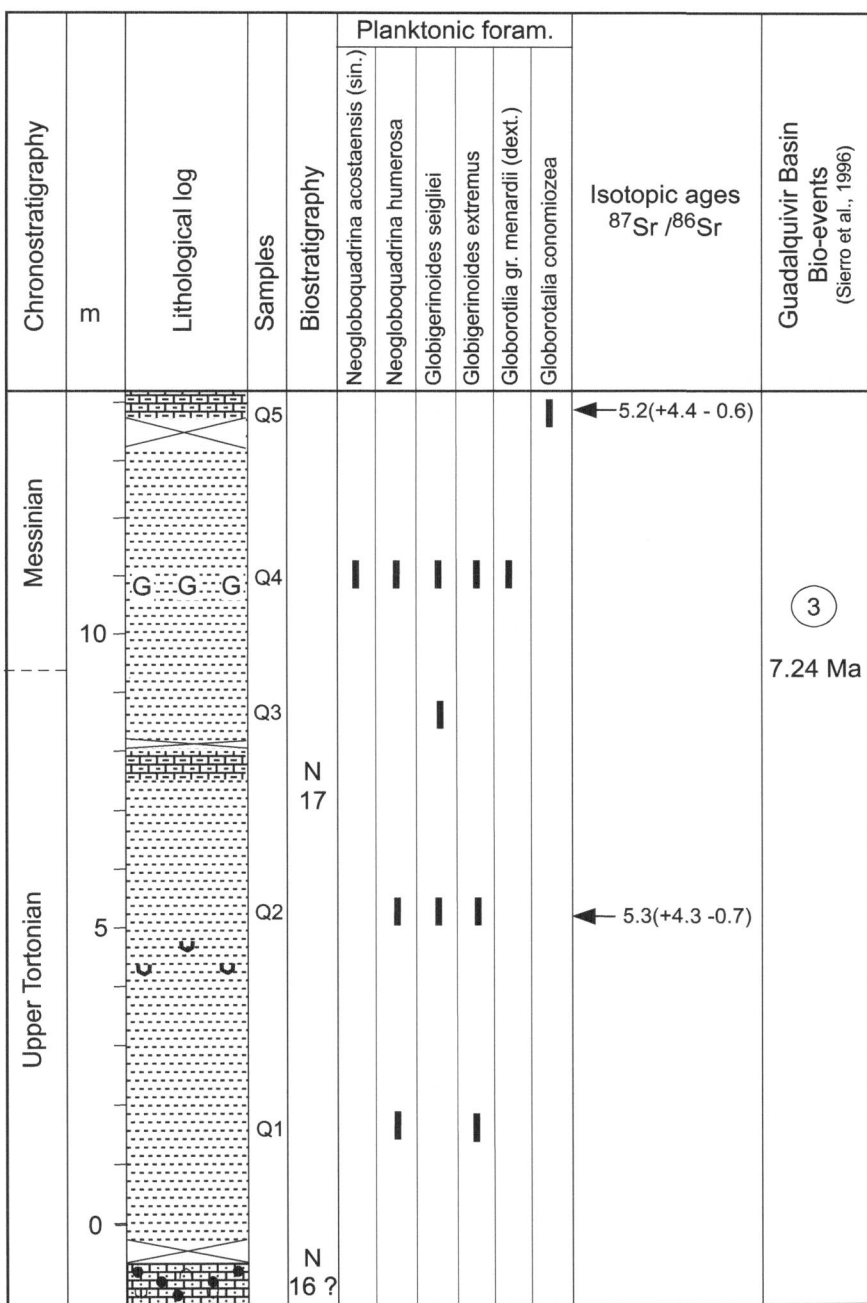

Fig. 33 The Quelfes stratigraphic section (Legoinha 2001, 2003)

Q3—*Globigerinoides bulloideus, Globigerinoides seigliei, Orbulina suturalis, Orbulina universa.*

Q4—*Globigerina bulloides, Globigerina concina, Globigerinoides bulloideus, Globigerinoides elongatus, Globigerinoides extremus, Globigerinoides seigliei, Globorotalia* gr. *menardii* (dext.), *Globorotalia miotumida, Neogloboquadrina acostaensis, Neogloboquadrina humerosa, Orbulina universa.*

Q5—*Globigerina bulloides, Globigerina druryi, Globigerinoides bulloideus, Globigerinoides triloba, Globorotalia conomiozea, Orbulina suturalis, Orbulina universa.*

Benthic foraminifera associations from Quelfes are characteristic of subtidal marine environments, with reasonable depth. Benthic foraminifera are common to abundant, and diversity is high. The highest abundance and diversity point to a Q4 peak depth. There are clear affinities with the higher layer of the Cacela Fábrica outcrop (Civis et al. 2000; Legoinha 2001, 2003):

Q1—*Ammonia, Bolivina, Bulimina, Cancris, Globobulimina, Globulina spinosa, Lagena, Lenticulina, Neoconorbina, Nodosaria badenensis, Nonion, Pullenia, Spiroplectamina, Uvigerina.*

Q2—*Bulimina, Cancris, Fursenkoina, Globulina spinosa, Heterolepa, Lenticulina, Nodosaria, Nonion, Pullenia, Reussela, Spiroplectamina, Textularia, Uvigerina.*

Q3—*Ammonia, Bulimina, Cancris, Cibicides, Dentalina, Elphidium, Globulina spinosa* (rara), *Guttulina* (rara), *Heterolepa, Lagena, Lenticulina, Nodosaria badenensis, Nonion, Textularia, Uvigerina*

Q4—*Ammonia, Asterigerina, Bulimina, Cancris, Cibicides, Elphidium, Globulina spinosa, Gyroidina, Heterolepa, Lenticulina, Neoconorbina, Nodosaria badenensis, Nodosaria hispida, Nonion, Pullenia, Reussela, Sphaeroidinella bulloides, Textularia, Uvigerina, Valvulineria bradyana.*

Q5—*Cibicides lobatulus, Cibicides* spp., *Elphidium* (abund.), *Fursenkoina, Lenticulina, Neoconorbina, Nodosaria, Planodiscorbis, Pullenia, Reussella, Sphaeroidinella bulloides, Textularia, Valvulineria bradyana.*

Isotopic $^{87}Sr/^{86}Sr$ determinations for Quelfes have given ages of 5.3 (+4.3,−0.7) Ma, 5.2 (+4.4,−1.1) Ma and 5.2 (+4.4,−0.6) Ma (Table 2 and Fig. 37).

These ages seem young when compared with the biostratigraphic indications. The study of planktonic foraminifera suggests an age of 7 (+1,−1) Ma for the sediments at Quelfes. It is therefore likely to be necessary to recalibrate the isotopic reference curve for this time interval. Quelfes section provides valuable information for this. It should be noted that glauconite sediments, correlative of Quelfes in the region of Luz de Tavira, gave K–Ar ages of 6.88 ± 0.4 Ma and 7.03 ± 0.4 Ma (Antunes et al. 1986c).

Event 3 from the Guadalquivir Basin is dated at 7.24 Ma, and event 2 at 7.35 Ma (F. Sierro, Univ. Salamanca, personal communication).

8.2.3 Mem Moniz Spongoliths

In Mem Moniz, there is an outcrop of spongoliths, overlying the Cretaceous, with few macrofossils, but with abundant microfauna (Figs. 28 and 37). At the stratification surfaces, bones and scales of fish (Clupeids? and Serranidae) can be found. The section was first described by Romariz et al. (1979a). Similar deposits are unknown elsewhere in Portugal, but are widely represented in the periphery of the Mediterranean, particularly in SE Spain.

The rock is essentially composed of spicules of sponges, studied in detail by Pisera et al. (2006). The association is rich and diverse, and comprises mainly desmosponjas Astrophorideas and including Lithistides. The Hexactinelides are relatively rare. The most common are *Eurylus* sp. and probably *Geodia* sp. Some forms, such as *Samus* sp. and *Alectona wallichii*, were identified in the fossil record for the first time. The association of sponges is typical of moderate depths, below the level of storm waves (~ 100 m or slightly deeper).

Romariz et al. (1979a) assigned the Mem Moniz deposits to the upper Burdigalian/lower Langhian (zones N8 to N9) based on planktonic foraminifera. Antunes et al. (1981) reported the presence of *Globorotalia acostaensis* (sin.) indicating the Tortonian (N16). Antunes et al. (1990b), given the preponderance of *Globigerina bulloides*, *Neogloboquadrina acostaensis* accompanied by *Globigerinoides bulloideus*, *Globigerina drury*, *Globigerina quinqueloba* and *Globigerinita glutinata*, indicated an age not earlier than N16.

Cachão (1995), based on calcareous nannofossils (among others *Helicosphera carteri*, *Recticulofenestra pseudoumbilica*, *Coccolithus pelagicus*, *Cyclococcolithus macintyrei*, *Sphenolithus abies*, *Discolithina multipora*; Antunes et al. 1981), assigned these sediments to the top of the Serravallian (CN5a). Also, the isotopic dating of shells of foraminifera indicated a $^{87}Sr/^{86}Sr$ age of 12.5 (+0.7,-1.7) Ma (Pais et al. 2000).

Legoinha (2001) identified *Globigerina angustiumbilicata*, *Globigerina bulloides*, *Globigerina concina*, *Globigerina drury*, *Globigerina falconensis*, *Globigerinella aequilateralis*, *Globigerinoides bulloideus*, *Globoquadrina baroemoenensis*, *Neogloboquadrina acostaensis* (sinist.), *Neogloboquadrina humerosa* (sinist.). There is little doubt that these indicate a Late Miocene age. Besides *Neogloboquadrina acostaensis*, the presence of *N. humerosa* indicates the deposits are upper Tortonian (upper part of zone N16 or N17). The association is older than the late Messinian (6 Ma), because at that time *N. acostaensis* changed from sinistral to dextral winding. The absence of keeled *Globorotalia* prevents greater accuracy of age assignment. The occurrence of *Globigerina drury*, whose last occurrence datum (LAD) is usually positioned in Serravallian (N14), could place some reservations; however, Sierro (1984) also notes its presence in the Upper Miocene and Lower Pliocene of the Guadalquivir basin. The association of planktonic foraminifera must correspond to zone N16 of the lower Tortonian (Antunes et al. 1981, 1990b, 1992b; Pais et al. 2000; Legoinha 2001). This parallels the chronology of the Mem Moniz spongoliths with similar events in the peri-Mediterranean area (Spain, Algeria and Italy).

Benthic foraminifera have also been recognized (Legoinha, 2001): *Uvigerina,*
Elphidium, Nonion, Bulimina, Bolivina, and Fursenkoina Cassidulina; Cibicides,
Giroydina, Planulina and *Trifarina* are scarce. The scarce occurrence of ostracods
is noted. There is a decrease of benthic foraminifera in the middle part of the
section and an increase at higher levels. *Nonion* predominates in the associations,
but in the upper sample there is a relative increase of *Uvigerina.* At the bottom of
the section, no ostracods have been found.

Civis (*in* Antunes et al. 1990b) indicates that among the benthic foraminifera
Nonion boueanum and *Ammonia* predominate, being associated to a lesser extent
with Buliminids, Bolivinids Uvigerinids. The association is interpreted as suggesting a shallow environment with plenty of organic matter in the sediment and a
likely lack of oxygen. This could be explained by the existence of an oxygen
minimum zone near the surface, or through semi-confinement.

Ostracoda include *Aurila zbyszewskii, Nonurocythereis seminulum* and *Carin-*
ocythereis galilea (Antunes et al. 1981, 1990b, 1992b).

To the west of the Mem Moniz outcrop, at the Tunes/Alvaledes railway station,
powdery white limestone and clay marl are exposed, containing ostracoda (Nascimento, in Antunes et al. 1990b): *Cytheridea neapolitana, Falunia rugosa,*
Mutilus cf. *labiatus, Nonurocythereis seminulum, Aurila zbyszewskii, Carinocy-*
thereis galilea, compatible with an early Tortonian age.

8.2.4 Galvana Conglomerate

The stratigraphic position of the spectacular conglomerate of Galvana (Faro), with
striated blocks that exceed 1 m across, and intercalated glauconitic silts (Antunes
et al. 1984), is unclear. The silts have a K–Ar age of 6.72 ± 0.17 Ma, although
glauconite must have been redeposited from the silts of Campina de Faro (Antunes
et al. 1984, 1990b, 1992b) (Figs. 28 and 37).

The conglomerates are closely related to tectonic instability resulting from
flexure of the southern Algarve. The K–Ar age is consistent with the Guadalquivir
Basin olistostroma, although this is some distance away. In a borehole drilled at
Campina de Faro, there are conglomerates overlying glauconitic silts that may
match the Galvana conglomerates. If this is the case, they are younger than suggested from the silts and may date from the Messinian-Zanclean.

8.3 Pliocene to Pleistocene

8.3.1 Ludo Formation

Miocene units in the Algarve are overlain almost exclusively by siliciclastic
deposits with variable thickness but becoming generally thicker to the east,
reaching a maximum thickness of 70 m in the region of Ludo-Montenegro, representing the bulk of the Pliocene and Pleistocene sedimentary deposits (Figs. 28,

34, 35, 36 and 37). The sedimentary depocentre was displaced eastward from the Quarteira Fault after the Late Miocene, when the Algarve's easternmost sub-basin evolved as a Guadalquivir Basin domain (Antunes et al. 1990b). In addition, the fracture system of the Miocene in a demi-graben pattern eastward from the Quarteira fault (Silva 1988) led to the genesis of a huge accommodation space to receive the sediment supplied from either fluvial or marine origin. In contrast, sedimentation during the Pliocene and Pleistocene was restricted to depressions in the emerged Miocene paleo-relief, mainly a karstic landscape, westward from Olhos de Água. For those reasons, the best localities for defining stratigraphic relationships between the various post-Miocene sedimentary facies are located near Faro (central Algarve).

According to the heavy mineral assemblage, and zircon in particular, Triassic and Cretaceous rock and sediment were the main detrital sources for the Pliocene and Pleistocene sedimentation (Veiga-Pires et al. 2007). Moreover, sands were successively remobilized, transported and deposited in several sedimentary environments, which when coupled with the very rare fossils makes it very difficult to distinguish and correlate sedimentary units.

In an attempt to establish stratigraphic relationships between siliciclastic sedimentary facies reflecting the environmental evolution of the Pliocene and Pleistocene in the Algarve region, the Ludo Formation, comprising four members, was defined (Moura and Boski 1994) (Figs. 34, 35 and 37), as observed in the 30-m-thick sequence of sedimentary deposits exposed in the Ludo area near Faro.

Montenegro Member and Falésia (Olhos de Água) Sands

Sedimentary facies include white medium sand rich in K-feldspar ($\sim 7\%$), and white greyish fine sand where muscovite reaches 1.5% of the sedimentary constituents. The ichnofacies and sedimentary structures that characterize this member are compatible with a shallow marine environment. The Montenegro member overlies the Miocene paleo-relief similarly to the Falésia sands (or Olhos de Água sands, Manuppella 1992). Therefore, the Falésia sands (described below) must be contemporaneous with the Montenegro member of the Ludo Formation yet displaying very different sedimentary facies. In addition, both the Falésia and Montenegro sands are bounded at the top by the Faro-Quarteira sands. These latter sands were probably deposited during the Pleistocene (Manuppella et al. 1987), and therefore the Montenegro and Falésia sands in all likelihood formed during the Pliocene (Figs. 28, 34, 35 and 37).

The Olhos de Água section was originally described by Romariz et al. (1979b). The succession begins with alternating sand layers and pelites overlain by yellow sand intermixed with cross-stratifications and white feldspathic sands. Locally, the yellow sands crop out in cliffs and are covered with red mudstones. The latter are superimposed either on the white sands or Quarteira sands (Plio-Pleistocene). On these white sands, there are beach sands and gravels containing debris eroded from aquatic vertebrates (fish, bones of cetaceans, teeth and bones of Sirenia). Most

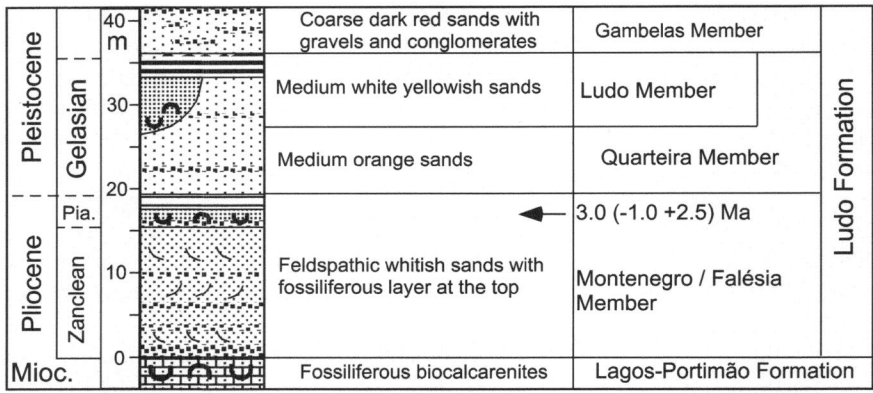

Pleistocene / Gelasian	40 m / 30 / 20	Coarse dark red sands with gravels and conglomerates	Gambelas Member	
		Medium white yellowish sands	Ludo Member	Ludo Formation
		Medium orange sands	Quarteira Member	
Pliocene / Pia. / Zanclean	Pia. / 10 / 0	◄ 3.0 (-1.0 +2.5) Ma		
		Feldspathic whitish sands with fossiliferous layer at the top	Montenegro / Falésia Member	
Mioc.	0	Fossiliferous biocalcarenites	Lagos-Portimão Formation	

Fig. 34 Synthetic log of the Plio-Pleistocene from the Algarve (modified from Moura *in* Terrinha et al. 2006)

marine fish usually have large dimensions (*Carcharocles megalodon, Isurus hastalis, Odontaspis* cf. *taurus, Hemipristis serra*, and *Carcharhinus* sp.). The degree of abrasion of the vertebrate remains suggests that they were rolled by the waves on shallow sandy bottoms near a beach.

The Sirenia (*Metaxytherium medium*) indicate coastal environments, rich in aquatic vegetation (algae, vascular plants). The presence of fish such as *Lates*, also common in freshwater, suggests (as do feldspathic sands) the existence of a river whose mouth was nearby. The presence of large crocodiles such as *Tomistoma* cf. *lusitanica* corroborates these views, and indicates sub-tropical conditions (Antunes 1979c; Antunes et al. 1981, 1990b, 1992b).

Just above this layer there are fine sands with poor microfaunal associations, and foraminifera including: *Elphidium crispum, Ammonia becarii, Nonion boueanum, Trifarina anulata, Rosalina* sp. and *Globigerina* sp. The top layer is a coarse biocalcarenite dominated by *Ostrea* gr. *lamellosa*, in addition to scarce pectinidae, including the rare *Paliollum (Lissochlamys) excisum*. The presence of molluscs indicates possibly a brackish environment (estuary).

The vertebrates are consistent with a post-Langhian chronology; they are certainly pre-Pliocene and match the Serravallian or the Tortonian. The molluscs are not chronologically characteristic. However, *Paliollum (Lissochlamys) excisum* is known both in the Upper Miocene and (mostly) in the Pliocene (Demarcq 1979) (common in the western Portuguese Upper Pliocene). The large bivalve Venerids (internal casts of *Pelecyora gigas*) disappeared from the Mediterranean region at ~3.0 Ma (Monegatti and Raffi 2001), suggesting a Late Miocene to early Piacenzian age for these deposits. Oyster shells yielded $^{87}Sr/^{86}Sr$ dates of 3.0 (+2.5,−1.0) Ma (Piacenzian, Late Pliocene). The vertebrate fossils could be reworked from the underlying Lagos-Portimão Formation.

A siltstone layer, located 2 m above these biocalcarenites, contains a magnetic polarity reversal that may correspond to the top of the Gauss event, 2.59 Ma

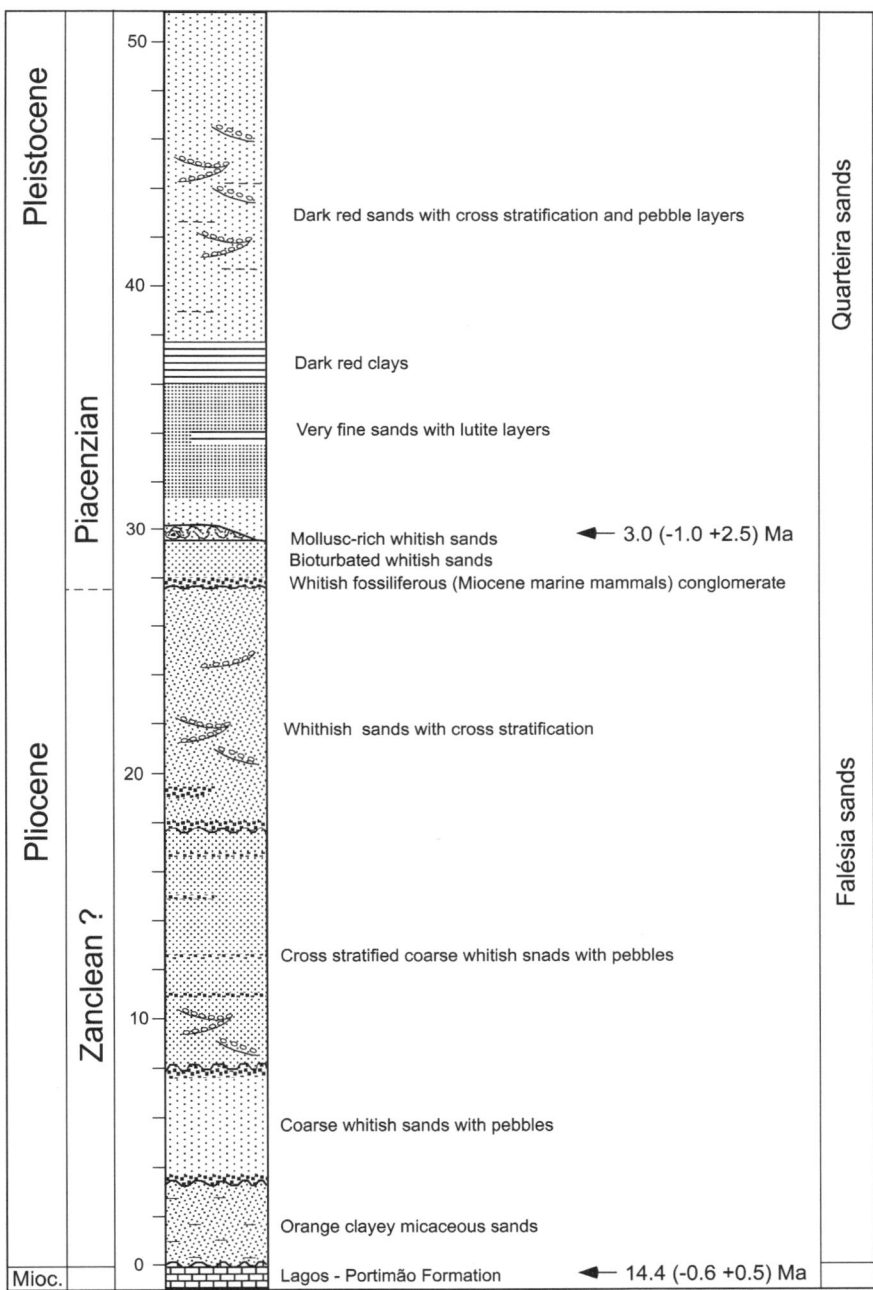

Fig. 35 The Falésia beach stratigraphic section (Pais et al. 2000)

(Moura 1998), in good agreement with the $^{87}Sr/^{86}Sr$ dates of 3.0 (+2.5,−1.0) Ma (Pais et al. 2000). Therefore, taking the updated lower limit to the Pleistocene (2.6 Ma) as proposed by the International Union of Quaternary Research (INQUA) and accepted in 2009 by the International Commission on Stratigraphy, the Faro-Quarteira sands are Pleistocene and the underlying Falesia sands (as well as the Montenegro sands) are Pliocene.

In Vale de Éguas, the presence of *Globigerinoides obliqus extremus* has allowed the Quarteira sands, overlying the Falésia sands, to be assigned to the transition fauna from warm to cold water by the end of the Pliocene (Aguirre and Pasini 1985; Harland et al. 1989).

The Falésia sands (Lower Pliocene) correspond to deposition in shallow coastal environment, in a demi-graben structured basin, after the Late Miocene (Manuppella, 1992). The sediment supply to the fluvial discharge was secured by the Cretaceous basement and volcanic-sedimentary complex and also Lower Jurassic sandstones and conglomerate units. It is likely that the Algarve mountains also contributed, albeit to a lesser extent, to feeding the river system, as did rounded grains inherited from the Mesozoic formations after several cycles of erosion and deposition (Terrinha et al. 2006).

At the top of the Ludo Formation, near Guia, an association of small mammals has been recorded; the predominance of *Oryctolagus cuniculus*, reptiles and amphibians (Antunes et al. 1989) indicates the presence of temporary pools of water. Although an age has not been accurately determined, these vertebrates must be fairly recent, and may be Holocene.

Faro-Quarteira Member

The term Faro-Quarteira sands was first used by Manuppella et al. (1987) as medium iron-rich feldspathic sands. However, the term was subsequently adopted by several authors to include almost all the post-Miocene siliciclastic sediments. This led to several cartographic and paleoenvironmental misinterpretations, as the Faro-Quarteira sands clearly differ from several other siliciclastic sediments with which they contact through erosional discontinuities. Therefore, in order to stratigraphically frame the Faro-Quarteira sands, they have been defined as a member of the Ludo Formation, constrained below by the Montenegro Member (and Falésia sands) and above by either the Ludo Member or the Gambelas Member (Figs. 28, 34, 35, 36 and 37).

The Faro-Quarteira Member (Faro-Quarteira sands) exhibits a remarkable sedimentological facies monotony. The deposits are very feldspathic, comprising medium sands, rarely coarse. Although being mainly massive, they may display horizontal or oblique stratification. The clayey matrix is probably responsible for the common appearance of joints resulting from differential compaction. Iron oxides (mainly hematite) in the clayey matrix confer on the deposits a characteristic reddish colour.

As referred to above, Mannuppella et al. (1987) assigned a Pleistocene age to the Faro-Quarteira sands. Their characteristics are compatible with deposition on a shallow continental shelf. An upper chronological limit to the Faro-Quarteira sands of 35,000–40,000 yr BP was reported by Chester and James (1995). However, it is not clear whether the considered deposits refer only to the Faro-Quarteira sands first defined by Manuppella et al. (1987) and therefore to this member of the Ludo Formation, or also include other siliciclastic deposits, such as the upper member of the Ludo Formation (Gambelas sands and pebbles).

Ludo Member

The Ludo member contacts with the underlying Faro-Quarteira Member through an erosional surface. It is composed of white, feldspathic, medium- and coarse-grained massive sand or, more rarely, displaying cross-stratification. Numerous linings of pulmonate gastropods occur within these sandy units. The Ludo sands are probably of fluvial origin (Figs. 28, 34, 36 and 37).

Gambelas Member

The Gambelas member contacts with all the previous members, as well as with Mesozoic and Cenozoic units, through erosional surfaces. This member is composed of very coarse and poorly-sorted gravelly sand with profuse oblique stratification. It is clay matrix-supported, and very rich in iron oxides, leading to a reddish colour. The sedimentary facies suggests a braided drainage network developed under arid climatic conditions. This scenario is in agreement with the chronological upper limit proposed by Chester and James (1991) (discussed above). Cold and arid conditions during the Last Glacial Period, together with a mean sea level lower than the present level, would have driven intense erosion processes. In addition, the present drainage channels incise all the Ludo Formation members, including the Gambelas, implying a later stage of channel-deepening. This could have happened during the Last Glacial Maximum (ca. 18 kyr BP) when the relative mean sea level was about 120–130 m below the present level (Figs. 28, 34, 35, 36 and 37).

8.3.2 Morgadinho and Algoz Deposits

This unit comprises thick sandy deposits associated (at the top) with marls, clays, lignite, lacustrine limestone and carbonate crusts. The deposits are poorly exposed, and most observations of them have been made during the opening of wells.

In Luz de Tavira and Morgadinho, these sands overlap the Cacela Formation and are overlain by the Gambelas sand and gravels as in Algoz (where the substrate is unknown).

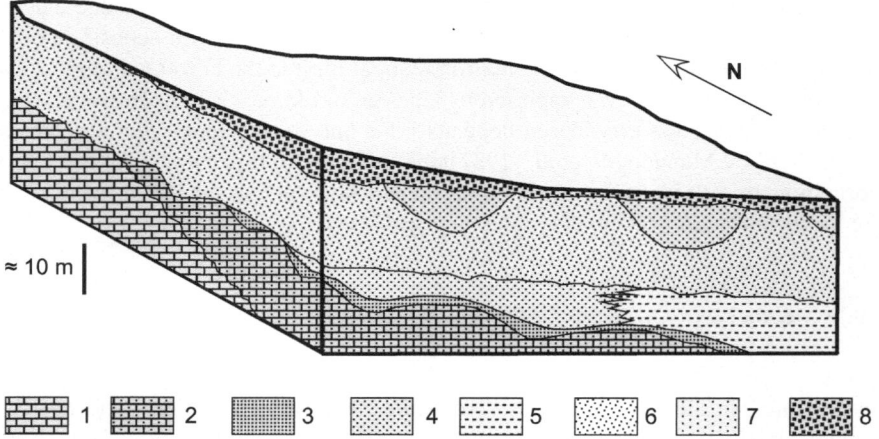

Fig. 36 Schematic block diagram showing the vertical and lateral distribution of the different members of the Pliocene–Pleistocene Ludo Formation, and their relationship with the underlying Mesozoic and Miocene units (scale approximate; adapted from Moura and Boski 1999 and Dias and Cabral 2002). *1* Mesozoic predominately carbonate rocks; *2* Lagos-Portimão sandy limestones (Miocene); *3* Cacela clayey silts (Miocene); *4* Monte Negro marine fine sands (Lower Pliocene); *5* Falésia marine sands (Lower Pliocene); *6* Faro-Quarteira fluvial and marine sands (Upper Pliocene); *7* Ludo fluvial sands (Pleistocene?); *8* Gambelas fluvial sands and gravels (Upper Pleistocene)

In Morgadinho, clays and lignite contain small mammals, fish, gastropods, freshwater ostracods, pollen and spores (Antunes et al. 1986a). The mammal fauna indicates an interval between the Late Pliocene and the Early-Middle Pleistocene, corresponding to mammal units NM17 NM20.

The Morgadinho deposits can be correlated with those of Algoz. The mammal fauna from Algoz (including *Eucladoceros* and *Hippopotamus*) is accurately dated to about 1 Ma (Biahrian, Early Pleistocene) (Zbyszewski 1950; Antunes et al. 1986a, b, c).

8.3.3 Paleoenvironmental Evolution

Based on the sedimentological facies, the geometric relationships between the post-Miocene siliciclastic deposits, and the few available chronological data, the following paleoenvironmental evolution during the Pliocene and Pleistocene in the Algarve region is proposed (Figs. 20, 21, 22, 23, 24, 25, 26 and 27).

During the Pliocene, a marine transgression led to the deposition of siliciclastics, whose geometry and sedimentary facies were controlled by the Miocene paleotopography and geographic position relative to the coastline. Falésia-Montenegro sands were deposited in coastal environments during the Pliocene. There was progradation of deltaic sedimentary bodies to the SSE; Falésia sands may

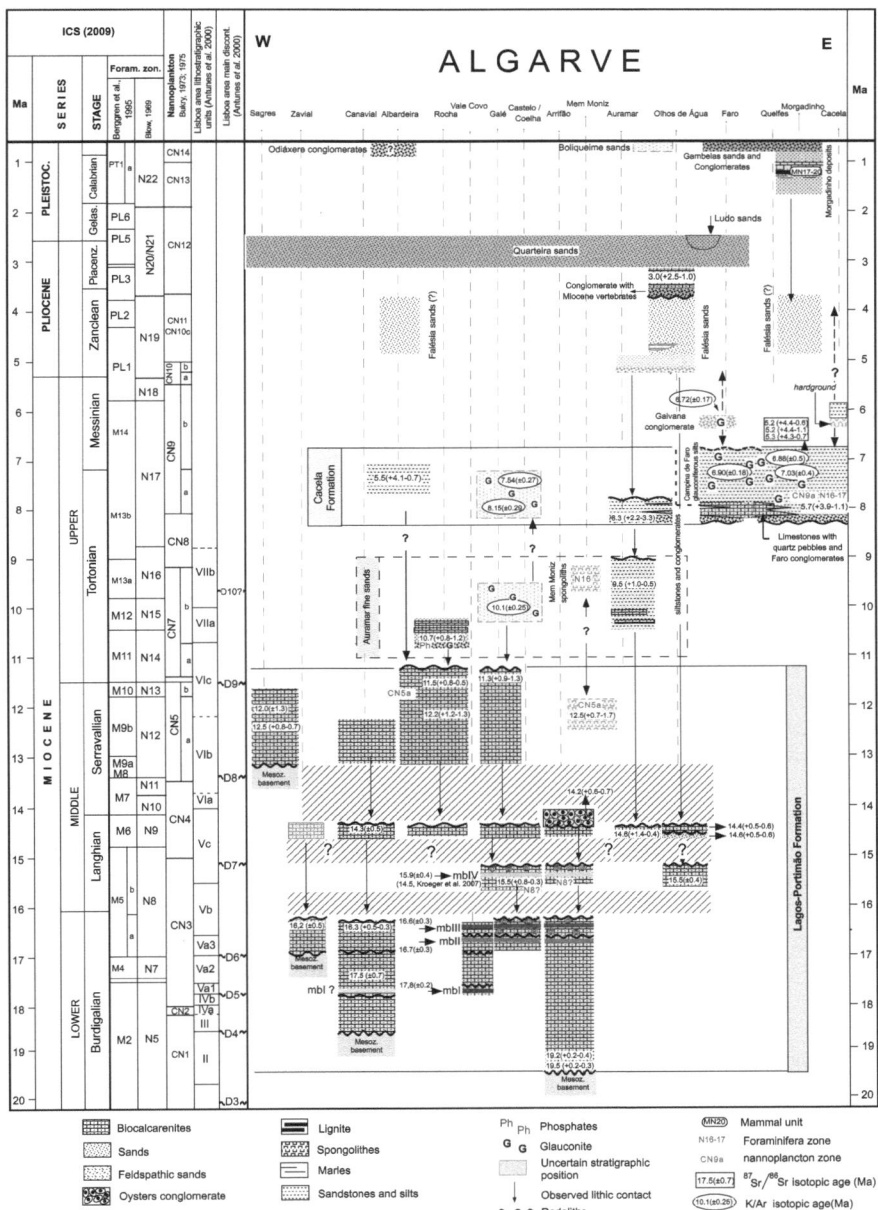

Fig. 37 Cenozoic of the Algarve (modified from: Pais et al. 2000; Legoinha 2001; Forst 2003; Kroeger et al. 2007)

correspond to the proximal portions of these transitional systems and the Montenegro sands to the most distal portions. Climatic conditions were warmer than today, and the mean sea level 50 m higher than at present in the Late Pliocene between 3.5 and 3.0 Ma (Haq et al. 1987). Both climatic conditions and sea level changed frequently during the Pleistocene.

The Quarteira sands were deposited on a shallow continental shelf, which extended at least 10 km further north than the existing shoreline. The association of foraminifera found within this unit is characteristic of marine and restricted sand bottom and is part of the biofacies transition that took place in the North Atlantic in the terminal Pliocene (Berggren and Olsson 1986; Poag and Miller 1986).

The Early Pleistocene coincided with a sharp cooling, accompanied by glacio-eustatic changes in sea level. Consequently, the river networks changed profoundly and the rivers became incised in the substrate. The Ludo sands indicate high sedimentation rates corroborated by the presence of casts of terrestrial gastropods. The Ludo sands became channelled in the Quarteira sands and had a discontinuous geographical distribution.

In the Middle to Late Pleistocene, a braided river network was established comprising channels with great lateral mobility, which were responsible for the deposition of the Gambelas sand and gravels. Several marine and fluvial terraces, and associated deposits such as the Odiáxere conglomerate, were formed. The presence in these sands of *Ruditapes decussata* and *Cardium* sp., and some quartz, quartzite and greywacke lithic implements, seems to indicate the Mousterian, which corresponds to the Riss-Würm interglacial and/or the initial part of the Würm glaciation (Upper Pleistocene). The stratigraphic position of these coarse layers had been allotted by Antunes et al. (1986a), who assigned these deposits as being younger than the Morgadinho deposits, which were assigned to the Middle Pleistocene. Moreover, a paleosoil occurring at the base of the above referred gravelly and sandy deposits of Gambelas should be contemporaneous with the Boliqueime deposits coeval with the Eemian warm climatic event.

The final Pleistocene sedimentary cover is geographically discontinuous, resulting from the coalescence of ancient alluvial fans. The pebble-sized clasts are of quartzite, greywacke and shale and are lithologically distinct from all older units (Moura and Boski 1999; Moura et al. 1998).

Acknowledgments This book is the result of the work developed under research projects funded by the Fundação para a Ciência e a Tecnologia (Ministério da Ciência e Ensino Superior) of Portugal: POCTI/32345/CTA/2000—Recognition of the Miocene of the distal region of the Lower Tagus Basin through a borehole with continous sampling; POCTI/CTA/38659/2001—The Tertiary of central-northern Portugal: basin analysis, stratigraphy and resources; PTDC/CTE-GIN/66283/2006—Paleoseismological Study of Active Faults in Mainland Portugal; and by the Ministerio de Ciencia e Innovación of Spain—CGL2009-11539/BTE—Registro paleontologico, sedimentologico e isotópico de cambios globales en ecosistemas de cuencas neogenas del Atlantico entre Lisboa (Portugal) y Tarfaya (Marruecos); CGL2008-0346—Tectonic morphology of the Guadiana river basin. Interaction between crustal deformation and drainage network.We acknowledge, for their assistance in various ways, the Research Centre on Science and Geological Engineering (CICEGe, FCT/UNL), the Marine Research Centre (CIMA-UALG), the IMAR—Marine and Environmental Research Centre (FCT/UC), the Geology Centre (UP) and Earth

Sciences Centre (UM), the Laboratório de Tectonofísica e Tecnónica Experimental/Instituto Dom Luís (LATTEX/IDL, FC/UL), and the Laboratório Nacional de Energia e Geologia (LNEG).

References

Afilhado A, Matias L, Shiobara H, Hirn A, Victor LM, Shimamura H (2008) From unthinned continent to ocean: the deep structure of the West Iberia passive continental margin at 38°N. Tectonophysics 458:9–50

Aguirre E (1997) The pliocene–pleistocene transition in the Iberian Peninsula. In: Van Couvering (ed) The Pleistocene boundary and the beginning of the quaternary. World and regional geology, vol 9. Cambridge University Press, Cambridge, pp 169–177

Aguirre E, Pasini G (1985) The pliocene–pleistocene boundary. Episodes 8(2):116–120

Alberdi MT, Antunes MT, Sondaar PY, Zbyszewski G (1978) Les hipparion du Portugal. Ciências da Terra 4:129–156

Alonso-Gavilán G, Armenteros I, Carballeira J, Corrochano A, Huerta P, Rodriguez JM (2004) Cuenca del Duero. In: Vera JA (ed) Geologia de España. SGE-IGME, Madrid, p 890

Alves TM, Gawthorpe RL, Hunt DH, Monteiro JH (2002) Jurassic tectono-sedimentary evolution of the northern Lusitanian basin (offshore Portugal). Marine Petrol Geol 19:727–754

Alves TM, Gawthorpe RL, Hunt DW, Monteiro JH (2003) Cenozoic tectono-sedimentary evolution of the western Iberian margin. Marine Geol 195:75–108

Alves TM, Cunha T, Moita C, Terrinha P, Monteiro JH, Manupella G (2011) A evolução de bacias sedimentares tipo-rift em margens continentais passivas: o exemplo da margem Ibérica Ocidental. In: Geologia de Portugal no contexto da Ibéria, Escolar Editora (in press)

Alves TM, Moita C, Cunha T, Ullnaess M, Myklebust R, Monteiro JH, Manupella G (2009) Diachronous evolution of Late Jurassic-Cretaceous continental rifting in the northeast Atlantic (West Iberian Margin): Tectonics 28(4):1–32

Alves TM, Moita C, Sandnes F, Cunha T, Monteiro JH, Pinheiro LM (2006) Mesozoic–Cenozoic evolution of North Atlantic continental-slope basins: the Peniche basin, western Iberian margin. AAPG Bull 90(1):31–60

Andeweg B, De Vicente G, Cloetingh S, Giner J, Muñoz-Martin A (1999) Local stress fields and intraplatedeformation of Iberia: variations in spatial and temporalinterplay of regional stress sources. Tectonophysics 305:153–164

Andrade MM (1944) Estudo polínico de algumas formações turfo-lignitosas portuguesas. Pub. Mus. Lab. Min. Geol. Fac. Ciências Univ. Porto, 37(2ª sér.):5–11

Andres I (1982) Estudo malacologico (Clase Bivalia) del Plioceno marino de Bonares (Huelva). Ph.D., Univ. Salamanca, p 410

Antunes MT (1964) Présence du genre Palaeotherium Cuv. (Equoidea, Mammalia) dans les argiles de Coja (Arganil). Considérations sur l'âge et l'extension des formations éocènes au Portugal. Rev. Fac. Ciên. Lisboa, 2ª sér. C, 13:103–122

Antunes MT (1967) Dépôts paléogènes de Côja: nouvelle données sur la Paléontologie et la Stratigraphie. Comparaisons avec. D'autres formations paléogènes. Rev. Fac. Ciênc. Lisboa, 2ª sér. C 15(1):69–111

Antunes MT (1975) Iberosuchus, crocodile Sebecosuchien nouveau, l'Éocène ibérique au Nord de la chaîne Centrale et origine du cañyon de Nazaré. Com Serv Geol Portugal 59:285–330

Antunes MT (1979a) Ensaio de síntese crítica acerca do Cretácico terminal e do Paleogénico de Portugal. Ciências da Terra 5:145–174

Antunes MT (1979b) "Hispanotherium fauna" in Iberian Middle Miocene, its importance and paleogeographical meaning. Ann Géol Pays Hell t hors sér I:19–26

Antunes MT (1979c) Vertebrados miocénicos de Olhos de Água (Algarve) interesse estratigráfico. Bol Mus Lab Min Geol Fac Ciências de Lisboa 16(1):343–352

Antunes MT (1983) Carta geológica de Portugal à escala de 1:50 000, Notícia explicativa da folha 39C-Alcácer do Sal. Serv. Geol. Portugal, Lisboa, p 21

Antunes MT (1984) Essai de synthèse sur les mammifères du Miocène du Portugal. Vol. d'hommage au géologue G. Zbyszewski, Ed. Recherche sur les Civilisations, Paris, pp 301–323

Antunes MT (1986a) Paralophiodon cf. leptorhynchum (tapiroidea, mammalia) à Vale Furado: contribution à la connaissance de l'Éocène au Portugal. Ciências da Terra 8:87–98

Antunes MT (1986b) Anoplotherium (Mammalia, Artiodactyla) et Geochelone (Reptilia, Testudines) à Côja: les vértebrés fossiles et I' Éocène supérieur au Portugal. Ciências da Terra 8:99–110

Antunes MT (1986c) lberosuchus et Pristichampsus, crocodiliens de l'Éocène - données complémentaires, discussion, distribution stratigraphique. Ciências da Terra 8:111–122

Antunes MT (1990a) The proboscidean data, age and palaleogeography: evidence from the Miocene of Lisbon. In: Lindsey EH et al (eds) European neogene mammal chronology. Plenum Press, New York, pp 253–262

Antunes MT (1990/91) O Homem da Gruta da Figueira Brava (ca. 30 000 BP). Contexto ecológico, alimentação, canibalismo. Mem Acad Ciências de Lisboa 31:487–536

Antunes MT (1995) On the Eocene Equid (Mammalia) from Feligueira Grande, Portugal, Paranchilophus lusitanicus (Ginsburg, 1965). Taxonomic status, stratigraphic and palaeogeographical meaning. Com Inst Geol Mineiro 81:57–72

Antunes MT (2000) Miocene mammals from Lisbon and geologic age. A showcase for marinecontinental correlations. Ciências da Terra 14:343–348

Antunes MT, Balbino AC (2004) The Carcharhiniformes (Chondrichthyes Neoselachii) from the Alvalade Basin (Portugal). Rev Esp Paleontologia 19(1):73–92

Antunes MT, Balbino AC (2006) Latest Miocene Myliobatids (Batoidei, Selachii) from the Alvalade Basin, Portugal. Cainozoic Res 4(1–2):41–49

Antunes MT, Broin F (1977) Cheirogasler sp (O Testudines, Fam Testudinidae, Geochelone sl) du Paléogène de Naia, Tondela el l'âge du gisement. Ciências da Terra 3:179–195

Antunes MT, Chevalier JP (1971) Notes sur la Géologie et la Paléontologie du Miocène de Lisbonne VII - Observations complémentaires sur les madréporaires et les faciès récifaux. Rev Fac Ciênc 2ª sér C 16(2):291–306

Antunes MT, Colin J-P (2003) Charophytes from Silveirinha (?Upper Paleocene—Lowermost Eocene) according to Janine Riveline. Ciências da Terra 15:77–82

Antunes M, Gaudant J (2003) Poissons du Paléogène inférieur de Silveirinha (Portugal). Ciências da Terra 15:101–102

Antunes MT, Ginsburg L (1983) Les rhinocérotidés du Miocène de Lisbonne. Systématique, écologie, paléobiogéographie, valeur stratigraphique. Ciências da Terra 7:17–98

Antunes MT, Jonet S (1970) Requins de l'Helvétien Supérieur et du Tortonien de Lisbonne. Rev Fac Ciências Lisboa, 2ª sér C 161:119–280

Antunes MT, Mein P (1977) Contributions à la Paléontologie du Miocène moyen continental du Bassin du Tage III—Mammifères - Póvoa de Santarém, Pero Filho et Chões (Secorio). Conclusions générales. Ciências da Terra 3:143–165

Antunes MT, Mein P (1979) Le gisement de Freiria de Rio Maior, Portugal, et sa faune de mammifères; nouvelle espèce de Rotundomys, conséquences stratigraphiques. Geobios 12(6):913–919

Antunes MT, Mein P (1981) Vertébrés du Miocène moyen de Amor (Leiria)/Importance stratigraphique. Ciências da Terra 6:169–188

Antunes MT, Mein P (1986) Petits mammifères du Burdigalien inférieur (Universidade Católica, Avenida do Uruguay). Ciências da Terra, vol 8. Lisboa, pp 123–138

Antunes MT, Mein P (1989) Petits mammifères du miocène terminal du Bassin de Alvalade (Portugal); comparaisons avec des faunes de l'Espagne et du Maghreb. Boll Soc Paleont Italiana 28(2–3):161–170

Antunes MT, Mein P (1992) Les plus anciens mammifères terrestres du Miocène marin de Lisbonne - le gisement du km10. Ciências da Terra 11:127–140

Antunes MT, Mein P (1995) Nouvelles données sur les petits mammifères du Miocène terminal du Bassin de Alvalade, Portugal. Com Inst Geol Mineiro 81:85–96

Antunes M, Pais J (1978) Notas sobre os depósitos de Taveiro: estratigrafia, paleontologia, idade, paleoecologia. Ciências da Terra 4:109–128

Antunes MT, Pais J (1984) Climate during Miocene in Portugal and its evolution. Paléob Contin 14(2):75–89

Antunes MT, Pais J (1992a) The Neogene of Portugal. Ciências da Terra, n° esp 2:13–24

Antunes MT, Pais J (1992b) 4 - Cenozóico. Estratigrafia, Algarve oriental. In: Carta geológica de Portugal na escala 1:200 000. Notícia explicativa da folha 8. Serviços Geológicos de Portugal, Lisboa, pp 64–67

Antunes MT, Pais J (1993) The neogene of Portugal. Ciências da Terra 12:7–22

Antunes MT, Russel D (1981) Le gisement de Silveirinha (Bas Mondego, Portugal): la plus ancienne faune de Vertébrés éocènes connues en Europe. C R Acad Sci Paris, Sér D 293:1099–1102

Antunes MT, Torquato JR (1969–70) Notes sur la Géologie et la Paléontologie du Miocène de Lisbonne. VI - La coupe de Quinta da Silvéria (Helvétien Vb et Vc): stratigraphie et évolution morphologique. Bol. Soc. Geol. Portugal, 17:1–30

Antunes MT, Ginsburg L, Torquato JR, Ubaldo M (1973) Age des couches à mammifères de la basse vallée du Tage (Portugal) et de la Loire moyenne (France). C R Acad Sci Paris, Sér D 277:2313–2316

Antunes MT, Bizon G, Nascimento A, Pais J (1981) Nouvelles données sur la datation des dépôts miocènes de l'Algarve (Portugal) et l'évolution géologique régionale. Ciências da Terra 6:153–168

Antunes MT, Ginsburg L, Mein P (1983) Mammifères miocènes de Azambujeira, niveau inférieur (Santarém, Portugal). Ciências da Terra 7:161–186

Antunes MT, Macedo CR, Pais J, Ferreira MP (1984) Datação K-Ar de glauconites do Miocénico superior de Galvanas e da Campina de Faro (Algarve oriental). Memórias e Notícias 98:1–7

Antunes MT, Azzaroli A, Faure M, Guérin C, Mein P (1986a) Mammifères pleistocènes de Algoz, en Algarve: une révision. Ciências da Terra 8:73–86

Antunes MT, Mein P, Nascimento A, Pais J (1986b) Le gisement pleistocène de Morgadinho, en Algarve. Ciências da Terra 8:9–22

Antunes MT, Odin G, Pais J (1986c) Ñges K-Ar de glauconies des environs de Luz de Tavira, Algarve. Ciências da Terra 8:22–30

Antunes MT, Calvo JP, Hoyos M, Morales J, Ordonez S, Pais J, Sese C (1987a) Ensayo de correlación entre el Neógeno de las areas de Madrid y Lisboa (Cuencas Alta y Baja del Rio Tajo). Com Serv Geol Portugal 73(1–2):85–102

Antunes M, Estravis C, Russell D (1987b) A new Condylarth (Mammalia) from the Early Eocene of Silveirinha, Portugal. Münchner Geowiss Abh (A) 10:219–224

Antunes MT, Crespo E, Mein P, Pais J, Teixeira JP (1989) Guia (Algarve), gisement de vertébrés quaternaires à caractère saisonnier. Ciências da Terra, vol 10, Lisboa, pp 97-106

Antunes MT, Civis J, González-Delgado JA, Pais J, Alonso-Gavilán G, Andrés I, Sierro FJ, Valle M, Nascimento A (1990a) The Late Serravallian-Early Tortonian marine sediments of the Tejo basin (Lisbon region). A sedimentological and paleontological approach IX RCMNS Congress, Barcelona, pp 27–29

Antunes MT, Civis J, Dabrio CJ, Sierro FJ, Glez-Delgado JA, Flores JA, Pais J, Valle M (1990b) El Neogeno del Algarve (Portugal) y de la cuenca del Guadalquivir (España). Actas de Paleontologia, Universidad de Salamanca 68:65–73

Antunes MT, Pais J, Legoinha P (1992a) Excursion A - Neogene deposits of Lisboa and Setúbal Peninsula. Ciências da Terra (UNL), n° esp 2:29–35

Antunes MT, Soulié-Marsh I, Mein P, Pais J (1992b) Le gisement de Aceisseira, Portugal (Miocène supérieur). Données complementaires sur Freiria de Rio Maior. Ciências da Terra 11:219–253

Antunes MT, Elderfield H, Legoinha P, Pais J (1995) Datações isotópicas com Sr do Miocénico do flanco Sul da serra da Arrábida. Com Inst Geol Min 81:73–78

Antunes MT, Civis J, Gonzalez-Delgado JA, Legoinha P, Nascimento A, Pais J (1996a) Miocene stable isotopes ($d^{18}O$, $d^{13}C$) biostratigraphy and environments in the southern limb of the Albufeira syncline (Setúbal Peninsula, Portugal). Geogaceta 21:21–24

Antunes MT, Legoinha P, Nascimento A, Pais J (1996b) The evolution of the lower Tejo basin (Lisbon and Setúbal Peninsula, Portugal from Lower to early Middle Miocene. Géol France 4:59–77

Antunes MT, Mein P, Pais J (1996c) Depósitos messinianos do sul de Portugal, mamíferos (incluindo rodentia e lagomorpha) e idades K-Ar. Ciências da Terra 8:55–64

Antunes MT, Casanovas ML, Cuesta MA, Checa L, Santafé JV, Agustí J (1997a) Eocene mammals from Iberian Peninsula. In: Aguilar JP, Legendre S, Michaux J (eds.) Actes dus Congr. BiochroM'97. Mem. Trav. E.P.H.E., Institut de Montepellier, vol 21, pp 337–352

Antunes MT, Elderfield H, Legoinha P, Pais J (1997b) The neogene of Algarve. Excursion 2 (Portuguese Part). Field Trip Guide, 2° Congreso RCANS, Salamanca, pp 37–55

Antunes MT, Elderfield H, Legoinha P, Pais J (1997c) A chronostratigraphical framework for the Miocene of the Lower Tejo Basin (Portugal); Depositional sequences, Biostratigraphy and Isotopic ages. Abstracts II Congress RCANS, Salamanca, pp 25–26

Antunes MT, Civis J, González-Delgado JA, Legoinha P, Nascimento A, Pais J (1998) Lower Miocene stable isotopes $d^{18}O$, $d^{13}C$, biostratigraphy and environments in the Foz da Fonte and Penedo sections (Setúbal Peninsula, Portugal). Geogaceta, Madrid 23:7–10

Antunes MT, Balbino AC, Cappetta H (1999a) Sélaciens du Miocène terminal du Bassin d'Alvalade (Portugal). Essai de synthèse. Ciências da Terra 13:115–129

Antunes MT, Elderfield H, Legoinha P, Nascimento A, Pais J (1999b) A Stratigraphic framework for the Miocene from the Lower Tejo Basin (Lisbon, Setúbal Peninsula, Portugal) depositional sequences, biostratigraphy and isotopic ages. Rev Soc Geol España 12(1):3–15

Antunes MT, Pais J, Balbino A, Mein P, Aguilar J–P (1999c) The Cristo Rei section (Lower Miocene). Distal fluviatile environments in a marine series, plants, vertebrates and other evidence, age. Ciências da Terra 13:141–155

Antunes MT, Legoinha P, Cunha PP, Pais J (2000) High resolution stratigraphy and Miocene facies correlation in Lisbon and Setúbal Peninsula (Lower Tejo basin, Portugal). Ciências da Terra 14:183–190

Armenteros I, Dabrio C, Alonso-Gavilán G, Jorquera A, Villalobos M (1986) Laminación y bioturbación en carbonatos lagunares: Interpretación genética (cuenca del Guadiana, Badajoz). Estudios Geol 42:271–280

Armenteros I, Corrochano A, Alonso-Gavilán G, Calvo J, Rodriguez JM (2002) Duero Basin. In: Gibson W, Moreno T (eds) The geology of Spain. The Geological Society of London, London, p 649

Azerêdo A, Carvalho AMG (1986) Novos elementos sobre o "Paleogénico" carbonatado dos arredores de Lisboa. Com Serv Geol Portugal 72(1–2):111–118

Azevêdo MT (1979) A Formação vermelha de Marco Furado (Península de Setúbal). Bol Soc Geol Portugal 21:153–162

Azevêdo MT (1982a) As formações quaternárias continentais da Península de Setúbal e sua passagem às formações litorais. Cad Lab Xeol Laxe Geologia 3:287–303

Azevêdo MT (1982b) O sinclinal de Albufeira, evolução pós-miocénica e reconstituição paleogeográfica. Ph.D., Univ. Lisboa, p 302

Azevêdo MT (1986) Reconstituição paleogeográfica do Tejo no Plio-quaternário. Actas do I Congresso do Tejo. Assoc. Amigos do Tejo. Lisboa

Azevêdo MT (1991) Essai de reconstitution paléogèographique du Bassin de Lisbonne au Paléogène. Mem Not 112:525–564

Azevêdo MT (1997) Depositional Architecture of the sedimentary infilling of the Pre-Tejo river in the upper Pliocene. ECSA Meeting 1997 Estuarine and Coastal Sciences Assoc. Inst. Oceanografia, FCUL, Lisboa

Azevêdo MT (2006) Interpretação das cinco fácies sedimentares do pré-Tejo no Pliocénico Médio da Península de Setúbal(Sul de Lisboa), VII Cong. Nac. Geologia, Livro de resumos 2:575–578

Azevêdo MT M, Cardoso JL (1986) Formações plio-quaternárias da Península de Setúbal—Guia da excursão da I Reunião do Quaternário Ibérico. G.T.P.E.Q. Setembro

Azevêdo MT, Pimentel NL (1995) Dados para a discussão da génese da Bacia do Tejo-Sado no Paleogénico. Actas do IV Cong. Nac. Geologia, Museu Labor. Mineral. Geol., Porto, Memória 4:897–902

Azevêdo MT, Cardoso JL, Penalva C, Zbyzsewsky G (1979) Contribuição para o conhecimento das indústrias líticas mais antigas do território português: as jazidas com "Pebble Culture" da formação de Belverde—Península de Setúbal. Setúbal Arqueológica 5:31–42

Balbino AC (1995) Seláceos (Pisces) do Miocénico terminal da Bacia de Alvalade (Portugal). Sistemática, ecologia, paleoambientes, comparação com faunas actuais. Ph.D., Univ. Évora, p 200

Balbino AC (1996) Sharks from the Midlle and early Upper Miocene from Lisbon, Portugal. A check-list. Comun. Inst. Geol. Mineiro 82:141–144

Barbosa B (1995) Alostratigrafia e litostratigrafia das unidades continentais da Bacia terciária do baixo tejo. Relações com o eustatismo e a tectónica. Ph.D., Univ. de Lisboa, p 253

Barbosa B, Reis RP (1989) Litostratigrafia e modelo deposicional dos sedimentos aluviais do Neogénico superior da Bacia do Tejo (Tomar-Lavre), Portugal. Com Serv Geol Portugal 75:89–97

Barbosa B, Reis RP (1996) Geometrias de enchimento, sistemas deposicionais e organização estratigráfica do pliocénico continental da Bacia Terciária do Baixo Tejo (Portugal). Com Serv Geol Portugal 82:51–86

Barra AP, Barbosa B, Martins AA, Reis RP (2000) Significado regional dos depósitos neogénicos continentais da área de Vila de Rei (Portugal Central). Ciências da Terra 14:163–170

Berggren W, Olsson R (1986) North Atlantic Mesozoic and Cenozoic palaeobiogeography. In: Vogt PR, Tucholke BE (eds) The geology of North América, vol M, pp 565–587

Berggren WA, Kent DV, Couvering JA van (1985) The neogene, 2. Geochronology and chronostratigraphy. In: Snelling NJ (ed) The chronology of the stratigraphic record. Geological Society London Memoirs, vol 10, pp 211–250

Bernardes CMA, Corrochano A (1987) A sedimentologia da 'Formação Arenitos e Argilas de Aveiro-Cretácico Superior, Bacia Ocidental Portuguesa. Geociências 2:9–26

Birot P (1949) Les surfaces d'érosion du Portugal central el septentrional. Rapport de la Commission pour la cartographie des surfaces d'aplanissemenl. Congr Intern Geographie, pp 9–116

Boski T, Moura D, Santos A, Delgado JA, Flores JA (1995) Evolução da bacia algarvia (Centro) durante o Neogénico. Mem Mus Lab Min Geol Univ Porto 4:47–51

Bourcart J, Zbyszewski G (1940) La faune de Cacela en Algarve (Portugal). Com Serv Geol Portugal 21:3–61

Brachert TC, Forst MH, Pais J, Legoinha P, Reijmer JJG (2003) Lowstand carbonates, highstands sandstones? Sedim Geology 155:1–12

Brébion P (1957) Le Mitra fusiformis des couches tortoniennes de Cacela est une espêce nouvelle: Mitra pereirai n. sp. Com Serv Geol Portugal 38:241–242

Brum da Silveira A (1990) Neotectónica e Sismotectónica da Região Vidigueira–Moura. MSc, Univ. Lisboa, p 204

Brum da Silveira A (2002) Neotectónica e sismotectónica de um sector do Alentejo oriental. Ph.D., Univ. Lisboa, p 339

Brum da Silveira A, Cabral J, Perea H, Ribeiro A (2009) Evidence for coupled reverse and normal active faulting in W Iberia The Vidigueira–Moura and Alqueva faults (SE Portugal). Tectonophysics 474:184–199

Cabral J (1995) Neotectónica em Portugal Continental. Memórias do Instituto Geológico e Mineiro 31:265

Cabral J, Ribeiro A (1988) Carta Neotectónica de Portugal Continental, escala 1:1000 000. Dep. Geol. Fac. Ciênc. Lisboa, Serv. Geol. Port., Gab. Prot. Seg. Nuclear, Ser. Geol. Port., Lisboa

Cabral J, Ribeiro A (1989) Incipient subduction along the West-Iberia continental margin. Abstracts, 28th International Geological Congress, Washington, D. C.; USA, 1/3:223

Cachão MA (1989) Contribuição para o estudo do pliocénico marinho português (sector Pombal-Marinha Grande)—Micropaleontologia, biostratigrafia. Provas Apt. Ped. Cap. Cient., Fac. Ciên. Univ, Lisboa, p 204

Cachão MA (1990) Posicionamento biostratigráfico da jazida pliocénica de Carnide. GAIA 2:11–16

Cachão MA (1995) Utilização de nanofósseis calcários em biostratigrafia, paleoceanografia e paleoecologia. Aplicações ao Neogénico do Algarve (Portugal) e do Mediterrâneo Ocidental (ODP 653) e à problemática de Coccolithus pelagicus. Ph.D., Univ. Lisboa, p 356

Callapez P (2003) Upper Paleocene-Early Eocene mollusks of Silveirinha (Figueira da Foz, West Central Portugal). Ciências da Terra 15:83–90

Calvo J, Daams R, Morales J, Lopez-Martínez N, Agusti J, Anadon P, Armenteros I, Cabrera L, Civis J, Corrochano A, Diaz-Molina M, Elizaga E, Hoyos M, Martin-Suarez E, Martínez J, Moissenet E, Muñoz A, Pérez-Garcia A, Pérez-Gonzalez A, Portero J, Robles F, Santisteban C, Torres T, Van der Meulen AJ, Vera J, Mein P (1993) Up-to-date Spanish continental Neogene synthesis and paleoclimatic interpretation. Rev Soc Geol Espana 6(3–4):29–40

Cardoso JL, Raposo L, Medeiros JP(1985) Novos elementos acerca do corte de Aldeia Nova e das indústrias líticas da região de Vila Real de Santo António. Actas da I Reunião do Quaternário Ibérico (Lisboa, 1985) 2:175–186

Carvalho AMG (1968) Contribuição para o conhecimento geológico da bacia terciária do Tejo. Mem Serv Geol Portugal 15:210

Carvalho AMG, Ribeiro A, Cabral J (1985) Evolução paleogeográfica da Bacia Cenozóica do Tejo-Sado. Bol Soc Geol Portugal 24:209–212

Casas-Sainz AM, De Vicente G (2009) On the tectonic origin of Iberian topography. Tectonophysics 474:214–235

Castro L (2006) Dinoflagelados e outros palinomorfos do Miocénico do sector distal da Bacia do Baixo-Tejo. Ph.D., Univ. Nova de Lisboa, p 380

Cebriá JM, López-Ruiz J, Doblas M, Martins LT, Munhá J (2003) Geochemistry of the Early Jurassic Messejana–Plasencia dyke (Portugal–Spain); Implications on the origin of the Central Atlantic Magmatic Province. J Petrol 44:547–568

Chavan A (1940) Les fossiles du Miocène supérieur de Cacela. Com Serv Geol Portugal 21:61–106

Chester D, James P (1991) Holocene alluviation in the Algarve, Southern Portugal: The case for an Anthropogenic cause. J Archaeol Sci 18:73–87

Chevalier JP, Nascimento A (1975) Notes sur la Géologie et la Paléontologie du Miocène de Lisbonne. XVI - Contribution à la connaissance des madréporaires et des faciès récifaux du Miocène inférieur. Bol Soc Geol Portugal 19(3):247–281

Choffat P (1906) Sur la tectonique de la chaîne de l'Arrábida entre les embouchures du Tage et du Sado. Bull Soc Géol France, sér 4e 6:237

Choffat P (1908) Essai sur la tectonique de la chaîne de l'Arrábida. Mem Comiss. Serv. Geol, Portugal, p 89

Choffat P (1950) Géologie du Cenozo du Portugal. Com Serv Geol Portugal 30(1):183

Civis J (ed) (2004) Cuencas Cenozoicas. Cap.6. Geología de España. SGE-IGME, pp 529–586

Civis J, Pais J, González-Delgado JA, Legoinha P (2000) Síntesis pleontológica del Tortoniense superior de Cacela (Algarve, Portugal). I Congresso Ibérico de Paleontología/XVI Jornadas de La Sociedad Española de Paleontologia, Livro de Resumos, pp 10–11)

Cloeting S, Burov E, Beekman F, Andeweg B, Andriessen P, Garcia-Castellanos D, De Vicente G, Vegas R (2002) Lithospheric folding in Iberia. Tectonics 21(5):1041–1067

Coelho AVP, Bravo MS (1983) Exemplo de vulcanismo tardio em Portugal. Rocha ígnea post-miocénico inferior (Figueira-Algarve). Ciências da Terra 7:99–114

Colin JP, Antunes MT (2003) Limnic ostracoda from Silveirinha, Portugal (? Late Paleocene-Lowermost Eocene). Ciências da Terra 15:91–100

Coney PJ, Muñoz JA, McClay KR, Evenchick CA (1996) Syntectonic burial and post-tectonic exhumation of the southern Pyrenees foreland fold-thrust belt. J Geo Soc London 153:9–16

Costa FP (1866) Gasterópodes dos depósitos terciários de Portugal. Com. Geol. Portugal, pp 5–252

Cotter JCB (1879) Fósseis das bacias Terciárias marinhas do Tejo, do Sado e do Algarve. Jorn. Sc. Math., Phys. e Naturaes 7(26):112–122

Cotter JCB (1956) O Miocénico marinho de Lisboa. Com Serv Geol Portugal 36(1):170

Cunha PP (1987a) Contribuição para o estudo sedimentológico dos depósitos terciarios da bacia de Sarzedas. A resposta sedimentar à modificação do contexto tectónico. Provas Apt. Pedag. Capac. Científica. Univ, Coimbra, p 147

Cunha PP (1987b) Evolução tectono-sedimentar terciária da região de Sarzedas (Portugal). Com Serv Geol Portugal 73(112):67–84

Cunha PP (1991) Estudo da paleodrenagem das "Arcoses de Côja" - (Portugal Central - Eocénico da Bacia Lusitaniana. 3° Congresso Nacional de Geologia (Resumos), Coimbra, p 100

Cunha PP (1992) Estratigrafia e sedimentologia dos depósitos do Cretácico Superior e Terciário de Portugal Central, a leste de Coimbra. Ph.D., Univ. Coimbra, p 262

Cunha PP (1996) Unidades litostratigráficas do Terciário da Beira-Baixa (Portugal). Com Inst Geo Mineiro 82:87–130

Cunha PP (1999a) Testemunhos geomorfológicos e sedimentológicos cenozóicos da transição da colmatação sedimentar para a gliptogénese, na área de Sarzedas-Vila Velha de Rodão (sector NE da Bacia do Baixo Tejo). Encontros de Geomorfologia, Coimbra, pp 61–68

Cunha PP (1999b) Unidades litostratigráficas do Terciário na região de Miranda do Corvo-Viseu (Bacia do Mondego - Portugal). Comun Inst Geol e Mineiro, Lisboa, vol 86, pp 143–196

Cunha PP (2000) Paleoalterações e cimentações nos depósitos continentais terciários de Portugal central: importância na interpretação de processos antigos. Ciências da Terra 14:145–154

Cunha PP, Martins AA (2004) Principais aspectos geomorfológicos de Portugal central, sua relação com o registo sedimentar e a importância do controlo tectónico. In Araújo MA, Gomes A (eds.) Geomorfologia do NW da Península Ibérica. FLUP, pp 155–182

Cunha PP, Pereira DI (2000) Evolução cenozóica da área de Longroiva-Vilariça (NE Portugal). Ciências da Terra 14:91–100

Cunha PP, Reis RP (1989) Principais ocorrências de paligorsquite, em depósitos de idade cretácica superior e terciária, em Portugal Central. 1° Reunião Luso-Espanhola de argilas, p 22

Cunha PP, Reis RP (1992) Establishment of unconformity-bounded sequences in the Cenozoic record of the western Iberian margin and syntesis of the tectonic and sedimentary evolution in central Portugal during Neogene. Abstracts of the First Congress RCANS, Lisboa, pp 33–35

Cunha PP, Reis RP (1995) Cretaceous sedimentary and tectonic evolution of the northern sector of the Lusitanian Basin. Cretaceous Res 16:155–170

Cunha PP, Barbosa B, Reis RP (1992) Proposal of synthesis concerning the Upper Pliocene infilling of the Lusitanian Basin, in the region between the parallels of Aveiro and Setúbal (Western central Portugal). Abstracts of the First Congress R.C.A.N.S., Lisboa, pp 37–42

Cunha PP, Barbosa BP, Reis RP (1993) Synthesis of the Piacenzian onshore record, between the Aveiro and Setúbal parallels (Western Portuguese margin). Ciências da Terra 12:35–43

Cunha PP, Pimentel NL, Pereira DI (2000) Assinatura tectono-sedimentar do auge da compressão bética em Portugal, A descontinuidade Valesiano terminal-Turoliano. Ciências da Terra 14:61–72

Cunha PP, Martins AA, Daveau S, Friend PF (2005) Tectonic control of the Tejo river fluvial incision during the late Cenozoic, in Ródão—central Portugal (Atlantic Iberian border). Geomorphology 64:271–298

Cunha PP, Martins AA, Huot S, Murray A, Raposo L (2008a) Dating the Tejo River lower terraces in the Ródão area (Portugal) to assess the role of tectonics and uplift. Geomorphology 102:43–54

Cunha PP, Martins AA, Pais J (2008b) O estudo do Cenozóico em Portugal continental—"estado da arte"e perspectivas futuras. In: Terra A (ed) conflitos e ordem. Livro Homen. Prof. AF Soares, Coimbra, pp 101–110

Cunha PP, Pais J, Legoinha P (2009) Evolução geológica de Portugal continental durante o Cenozóico-sedimentação aluvial e marinha numa margem continental passiva (Ibéria ocidental). 6° Simposio sobre el Margen Ibérico Atlántico MIA09, pp 11–20

Dabrio CJ, González Delgado JA, Armenteros I, Civis J, Pais J, Alonso Gavilán G, Legoinha P (2008) Facies changes and paleogeographical implications in the Serravallian of the Lagos-Portimão Formation (Praia da Rocha, southern Portugal). Geo-Temas 10:132–134

Daveau S (1985) Critères géomorphologiques de déformations tectoniques récentes dans les montagnes de schistes de la Cordilheira Central (Portugal). Bull Ass française l'étude du Quaternaire 4:229–238

Daveau S, Birot P, Ribeiro O (1985–86) Les bassins de Lousã et Arganil. Recherches géomorfologiques et sédimentologiques sur le massif ancien et sa couverture à l'Est de Coimbra. Mem Centro de Est Geog 8(I/II):450

De Broin FL (2003) Neochelys sp. (Chelonii, Erymnochelyinae), from Silveirinha, early Eocene, Portugal. Ciências da Terra 15:117–132

De Vicente G (ed) (2004) Estructura alpine del Antepaís Ibérico. In: Vera JA (ed) Geologia de España, SGE-IGME, Madrid, pp 587–634

De Vicente G, Vegas R, Cloetingh S, Muñoz A, Elorza FJ, Sokoutis D, Alvarez J, Olaiz A (2005) The Cenozoic constrictive deformation of Iberia. Thrust Belts and Foreland Basins, International Meeting, Rueil-Malmaison, Abstracts vol, pp 117–119

De Vicente G, Cloetingh S, Muñoz-Martín A, Olaiz A, Stich D, Vegas R, Galindo-Zaldivar J, Fernández-Lozano J (2008) Inversion of moment tensor focal mechanisms for active stresses around microcontinent Iberia: tectonic implications. Tectonics 27:1–22

De Vicente G, Vegas R (2009) Large-scale distributed deformation controlledtopography along the western Africa–Eurasia limit: tectonic constrains. Tectonophysics 474:124–143

De Vicente G, Cloetingh S, Van Wees JD, Cunha PP (2011) Tectonic classification of Cenozoic Iberian foreland basins. Tectonophysics 502(1–2):38–61

Demarcq G (1979) Éssais de synthèse biostratigraphique sur les pectinidés du Néogène mediterranéen. Ann Géol Pays Hellén t hors série 1:305–307

Dewey JF, Helman ML, Turco E, Hutton DHW, Knott SD (1989) Kinematics of the western Mediterranean. In: Coward MP, Dietrich D, Park RG (eds) Alpine tectonics. Geological society. Special Publication, London, pp 265–283

Dias RP (2001) Neotectónica da Região do Algarve. Ph.D., Fac.Ciências, Univ. Lisboa, p 369

Dias RP, Cabral J (2002) Interpretation of recent structures in an area of cryptokarst evolution–Neotectonics versus subsidence genesis. Geodinamica Acta 15(4):233–248

Dias RP, Pais J (2009) Homogeneização da cartografia geológica do Cenozóico da Área Metropolitana de Lisboa. Com Geol 96:39–50

Dias RP, Pais J, Barbosa B, Piçarra J (2006) Nova interpretação dos calcários do Cano-Casa Branca (nota preliminar). In: Mirão J, Balbino A (Coord.) Livro de Resumos, VII Cong. Nac. Geol., Univ. Évora 2:629–632

Dias RP, Barbosa B, Pais J, Azerêdo A, Pimentel N, Cabral MC (2009a) Cenozóico. In Carta Geológica de Portugal na escala 1:50000, Notícia Explicativa da Folha 32-C Avis. Unid. Geol. Cart. Geol., Lab. Nac. Energ. Geol., pp 12–17

Dias RP, Barbosa B, Pais J, Azerêdo A, Pimentel N, Cabral MC (2009b) Alterações Pedogenéticas de Idade Indeterminada (Paleogénico-Quaternário). In Carta Geológica de Portugal na escala 1:50000,Notícia Explicativa da Folha 32-C Avis, Unid. Geol. Cart. Geol., Lab. Nac. Energ. Geol., pp 22–27

Dinis J, Rey J, Cunha PP, Callapez P, Reis RP (2008) Stratigraphy and allogenic controls of the western Portugal Cretaceous: an updated synthesis. Cretaceous Res 29:772–780

Diniz F (1984) Apports de la palynology à la connaissance du Pliocène portugais. Rio Maior : un basin de référence pour l'histoire de la flore, de la végétation et du climat de la façade atlantique de l'Europe meridionale. Ph.D., Univ. Sc. Techn. Languedoc, Montpellier, p 230

Diniz F (2001) Aspects of the Plio-Quaternary Transition in Rio Maior: Pollen Records, Vegetation and Climate/Actas V Reunião do Quaternário Ibérico—I Congresso do Quaternário de Países de Línguas Ibéricas, pp 109–112

Dollfus G, Cotter JCB, Gomes JP (1903–1904) Mollusques Tertiaires du Portugal/Planches de Céphalopodes, Gastéropodes et Pélécypodes laissées par F.A.Pereira da Costa/accompagnées d'une explication sommaire et d'une esquisse géologique. Imprimerie de l'Académie Royale des Sciences, Lisboa, p 44

Dowsset HJ, Barron JA, Poore RZ (1996) Middle Pliocene sea surface temperature. A global reconstitution. Marine Micropal 27:13–25

Driscoll NW, Hogg JR, Christie-Blick N, Karner GD (1995) Extensional tectonics in the Jeanne d'Arc Basin, offshore Newfoundland: Implications for the timing of break-up between Grand Banks and Iberia. In: Scrutton RA, Stoker MS, Shimmield GB, Tudhope AW (eds) The tectonics, sedimentation and palaeoceanography of the North Atlantic region. Geol Soc Sp Publ. London, 90:1–28

Duran JJ, Garzón G, García de Domingo A, Muñoz P, Ortega J, Soria JM (2005) Génesis y evolución de lagunas endorreicas en superficies colgadas por abandono y degradación de una red de drenaje previa: el caso de la Albuera, Badajoz. Geogaceta 38:255–258

Escuder-Viruete J, Carbonell R, Jurado MJ, Martí D, Pérez-Estaún A (2001) A two dimensional geostatistical modelling and prediction of the fracturation in the Albala granitic pluton, SW Iberian Massif, Spain. J Struct Geol 23:2011–2023

Estevens M (2005) Mamíferos marinhos do Neogénico de Portugal. Ph.D., Univ. Nova de Lisboa, p 369

Estevens M, Antunes MT (2003) Fragmentary remains of Odontocetes (Cetacea, mammalia) from the Miocene of the lower Tejo Basin (Portugal). Rev Esp Paleont 19(1):93–108

Estravís C (1992) Estudo dos mamíferos do Eocénico inferior de Silveirinha (Baixo Mondego). Ph.D., Univ. Nova Lisboa, p 254

Estravís C (2000) Nuevos mamíferos del Eocene Inferior de Silveirinha (Baixo Mondego, Portugal). Col Paleont 51:281–311

Fauquette S, Suc JP, Guiot J, Diniz F, Feddi N, Zheng Z, Bessais E, Drivaliari A (1999) Climate and biomes in the West Mediterranean area during the Pliocene. Palaeog Palaeoclim Palaeoecol 152:15–36

Fernandez-Lozano J, Sokoutis D, Willingshofer J, Cloetingh S, G. De Vicente (2011) Cenozoic deformation of Iberia: a model for intraplate mountain building and basin development based on analogue modeling. Tectonics 30:26 [TC1001]

Ferreira OV (1951) Os pectinídeos do Miocénico do Algarve. Com Serv Geol Portugal 32:153–180

Ferreira AB (1971) O rebordo ocidental da Meseta e a depressão tectónica da Longroiva. Finisterra 6:196–217

Ferreira AB (1978) Planaltos e montanhas do Norte da Beira Estudo de geomorfologia. Mem Centro Estudos Geográficos 4:374

Ferreira AB (1991) Neotectonics in Northern Portugal. A geomorphological approach. 2. Geomorph N F sup.-Bd 82:73–85

Ferreira AB (2005) Formas do relevo e dinâmica geomorfológica. In: Medeiros (ed.) Geografia de Portugal—O Ambiente Físico, Círculo de Leitores 1:53–256

Fonseca B (1977) Notes sur la Géologie et la paléontologie du Miocène de Lisbonne XVIII– Coupe de Palença, rive gauche du Tage: stratigraphie et micropaléontologie (Coc- colithophoridés). Ciências da Terra 3:61–99

Forst MH (2003) Zur Karbonatsedimentologie, Biofazies und sequenzstratigraphischen Archi- tektur eines fossilen Hochenergie-Schelfs aus dem Neogen der Algarve (Miozän, Südpor- tugal). Ph.D., Johannes Gutenberg-Universität in Mainz, p 221

Forst MH, Brachert TC, Pais J (2000) High-resolution correlation of coastal cliff sections in the Lagos-Portimão Formation (Lower—Middle Miocene, central Algarve, Portugal). Ciências da Terra 14:289–296

Freneix S (1957) Lamellibranches nouveaux du Miocene du Portugal. Com Serv Geol Portugal 38:227–239

Friend PF, Dabrio CJ (eds) (1996) Tertiary basins of Spain. The stratigraphic record of crustal kinematics.World and regional geology series, vol 6. Cambridge University Press, Cambridge, p 400

Garzón G (2005) Geomorfología y paisaje extremeño. In: Barco M, Martínez-Flores (eds) Patrimonio Geológico de Extremadura.. Junta de Extremadura, 35–70

Gaudant J (1977) Contributions à la Paléontologie du Miocène moyen continental du Bassin du Tage II—Observations sur les dents pharyngiennes de poissons cyprinidés - Póvoa de Santarém. Ciências da Terra 3:129–141

Gibbons W, Moreno MT (eds) (2002) The geology of Spain. London geological society, London, p 649

Gonçalves F, Antunes MT (1992) Carta geológica de Portugal na escala 1:50 000. Notícia explicativa da folha 39-B Torrão. Serv Geol Portugal, p 86

González-Delgado JA, Sierro FJ, Civis J (1993) Cambios globales del nivel del mar y concentraciones de megafaunas marinas someras en el Neógeno del Oeste de la Península Ibérica. Comunicaciones de las IX Jornadas de Paleontología, Málaga, pp 33–36

González-Delgado JA, Andrés I, Sierro FJ (1995) Late Neogene Molluscan faunas from NE Atlantic (Portugal, Spain, Morocco). Geobios 28:459–471

Haq BU, Hardenbol J, Vail PR (1987) Chronology of fluctuating sea levels since the Triassic. Science 235:1156–1166

Haq BU, Hardenbol J, Vail PR (1988) Mesozoic and Cenozoic chronostratigraphy and cycles of sea-level change. In: Wilous C et al (eds) Sea-level changes. An integrated approach, vol 42. Soc. Econ. Paleo. Min. Sp. Pub, pp 39–44

Harland WB, Armstrong RL, Cox A, Craig LE, Smith AG, Smith DG (1989) Geological time scale. Cambridge University Press, Cambridge, p 263

Hernández Pacheco F (1960) El Terciario Continental de Extremadura. Real Soc Esp Hist Nat 63:2

Hernández-Pacheco F, Crusafont-Pairó M (1960) Primera caracterización paleontológica del Terciario de Extremadura. Bol Real Soc Esp Hist Nat Sec Geológica 63:275–282

ICS (2009) International stratigraphic chart. Int Comm Stratigraphy, IUGS

Juarez MT, Lowrie W, Osete ML, Melendez G (1998) Evidence of widespread Cretaceous remagnetization in the Iberian Range and its relation with the rotation of Iberia. Earth Planet Sci Lett 160:729–743

Klitgord KD, Schouten H (1986) Plate kinematics of the central Atlantic. In: Vogt PR, Tucholke BE (eds) The geology of North America, vol. M, The Western North Atlantic region. Geological Society of America, Boulder, pp 351–378

Kroeger KF, Reuter M, Forst MH, Breisig S, Hartmann G, Brachert TC (2007) Eustasy and sea water Sr composition: application to high-resolution Sr-isotope stratigraphy of Miocene shallow-water carbonates. Sedimentology 54:565–585

Kullberg JC (2000) Evolução tectónica mesozóica da Bacia Lusitaniana. Ph.D., Univ. Nova Lisboa, p 361

Kullberg JC, Manuppella G, Pais J (1992) Aspectos da tectónica alpina no Algarve. Ciências da Terra 11:293–302

Kullberg MC, Kullberg JC, Terrinha P (2000) Tectónica da Cadeia da Arrábida. In Tectónica das regiões de Sintra e Arrábida, Mem. Geociências. Univ. Lisboa 2:35–84

Kullberg JC, Rocha RB, Soares AF, Rey J, Terrinha P, Callapez P, Martins L (2006a) A Bacia Lusitaniana: Estratigrafia, Paleogeografia e Tectónica. In: Dias R, Araújo A, Terrinha P, Kullberg JC (eds) Geologia de Portugal no contexto da Ibéria. Univ. Évora, Évora, pp 317–368

Kullberg JC, Terrinha P, Pais J, Reis RP, Legoinha P (2006b) Arrábida e Sintra: dois exemplos de tectónica pós-rifting da Bacia Lusitaniana. In: Dias R, Araújo A, Terrinha P, Kullberg JC (eds) Geologia de Portugal no contexto da Ibéria. Univ Évora, Évora, pp 369–396

Kullberg JC, Rocha RB, Soares AF, Rey J, Terrinha P, Azerêdo AC, Callapez P, Duarte LV, Kullberg MC, Martins L, Miranda JR, Alves C, Mata J, Madeira J, Mateus O, Moreira M, Nogueira CR (2011a) A Bacia Lusitaniana: Estratigrafia, Paleogeografia e Tectónica. In: Geologia de Portugal no contexto da Ibéria, Escolar Editora (in press)

Kullberg JC, Rocha RB, Soares AF, Rey J, Terrinha P, Azerêdo AC, Callapez P, Duarte LV, Kullberg MC, Martins L, Miranda JR, Alves C, Mata J, Madeira J, Mateus O, Moreira, M, Nogueira CR (2011b) A Bacia Lusitaniana: Estratigrafia, Paleogeografia e Tectónica. In: Dias R, Araújo A, Terrinha P, Kullberg JC (eds) Geologia de Portugal no contexto da Ibéria, Escolar Ed (in press)

Lanaja JM (1987) Contribución de la exploración petrolífera al conocimiento de la Geología de España. Instituto Geológico y Minero de España, Madrid, p 465

Lauriat-Rage A, Brébion P, Cahuzac B, Chaix C, Ducasse O, Ginsburg L, Janin M-C, Lozouet P, Margerel J, Nascimento A, Pais J, Poignant A, Pouyet S, Roman J (1993) Palaeontological data about the climatic trends from Chattian to present along the Northeastern Atlantic frontage. Ciências da Terra 12:167–179

Legoinha P (2001) Biostratigrafia de foraminíferos do Miocénico de Portugal. Ph.D., Univ. Nova, Lisboa, p 238

Legoinha P (2003) Upper Miocene planktonic foraminifera from Algarve. Chronostratigraphical implications. Ciências da Terra 15:199–208

Legoinha P (2008) The Serravallian-Tortonian boundary in the Lower Tejo Basin (Portugal) and the new GSSP of the Tortonian stage. e-Terra 6(1):10

Legoinha P, Sousa L, Pais J, Ferreira J, Amado AR (2004) Miocene lithological, foraminiferal and palynological data from the Belverde borehole (Portugal). Rev Esp Paleontologia 19(2):243–250

LNEG (2010) Carta geológica de Portugal na escla de 1:1 000 000

Lopes F (2002) Análise tectono-sedimentar do Cenozóico da Margem Algarvia. Ph.D., Univ. Coimbra, p 593

Lopes FC, Cunha PP (2007) Tectono-sedimentary phases of the latest Cretaceous and Cainozoic compressive evolution of the Algarve margin (southern Portugal). Chapter 6 In: Nichols GJ, Williams EA, Paola C (eds) Sedimentary processes, environments and basins—a tribute to Peter Friend, IAS Special Publication 38. Wiley-Blackwell, Oxford, p 642

Lopes F, Cunha PP, Le Gall B (2006) Cenozoic seismic stratigraphy and tectonic evolution of the Algarve margin. Mar Geol 231:1–36

Lopes FC, Cunha PP, Le Gall B (2008a) Cenozoic salt tectonics in the Algarve Margin. Evidence from multichannel seismic data. Abstract Books. 4th. Topo Europe Meeting. El Escorial (Madrid), 5–8 Outubro

Lopes FC, Cunha PP, Le Gall B, Mendes-Victor LA (2008b) Tectónica Salífera Cenozóica na Margem Algarvia. In: Callapez PM, Rocha RB, Marques JL, Cunha LS, Dinis PM (eds) A Terra, Conflitos e Ordem, uma homenagem a António Ferreira Soares. MMGUC Publishing, Coimbra, pp 231–242

Malod JA (1989) Iberian kinematics: a rolling stone story. Annales Geophysicae, vol 26, sp. issue. Gauthier-Villars, Paris

Malod JA, Mauffret A (1990) Iberian plate motion during the Mesozoic. In: Boillot G, Fontbote JM (eds) Alpine evolution of Iberia and its continental margins. Tectonophysics 184:261–278

Manuppella G (Coord.) (1999) Carta geológica de Portugal na escala 1:50 000, Notícia explicativa da folha 39-B (Setúbal). Inst Geol, Mineiro, p 143

Manuppella G (1992) Carta geológica da região do Algarve, escala 1/100 000. Notícia explicativa da da Carta Geológica da região do Algarve. Serv Geol Portugal, Lisboa, p 15

Manuppella G, Ramalho M, Antunes MT, Pais J (1987) Carta geológica de Portugal na escala 1/50 000. Notícia explicativa da folha 53-A, Faro. Serviços Geológicos de Portugal, Lisboa, p 52

Martín Velázquez S, Elorza FJ (2007) Deformación cenozoica de la litosfera Ibérica: Sistema Central y cuencas del Duero y Tajo. Geogaceta 42:11–14

Martins AA, Cunha PP (2009) Terraços do rio Tejo em Portugal, sua importância na interpretação da evolução da paisagem e da ocupação humana. In: Arqueologia do Vale do Tejo, CPGP, Lisboa, pp 163–176

Martins A, Barbosa B, Reis RP (1998) Os Conglomerados de Rio de Moinhos (Abrantes–Portugal Central) Actas V Congresso Nacional de Geologia. Com Inst Geol Mineiro 84(1):A142–A144

Martins LT, Madeira J, Youbi N, Munhá J, Mata J, Kerrich R (2008) Rift-related magmatism of the Central Atlantic magmatic province in Algarve, Southern Portugal. Lithos 101:102–124

Martins AA, Cunha PP, Huot S, Murray A, Buylaert JP (2009a) Geomorphological correlation of the tectonically displaced Tejo River terraces (Gavião-Chamusca area, central Portugal) supported by luminescence dating. Quat Int 199:75–91

Martins AA, Cunha PP, Matos J, Guiomar N (2009b) Quantificação da incisão do rio Tejo no sector entre Gavião e Chamusca, usando os terraços fluviais como referências geomorfológicas. Public. Assoc. Portuguesa de Geomorfólogos 6:83–86

Martins AA, Cunha PP, Buylaert J-P, Huot S, Murray AS, Dinis P, Stokes M (2010a) K-Feldspar IRSL dating of a Pleistocene river terrace staircase sequence of the Lower Tejo River (Portugal, western Iberia). Quat Geochrono 5(2–3):176–180. doi:10.1016/j.quageo.2009.06.004

Martins AA, Cunha PP, Rosina P, Osterbeck L, Cura S, Grimaldi S, Gomes J, Buylaert J-P, Murray AS, Matos J (2010b) Geoarcheology of Pleistocene open-air sites in the Vila Nova da Barquinha-Santa Cita area (Lower Tejo River basin, central Portugal). Proc Geol Assoc 1212:128–140. doi:10.1016/j.pgeola.2010.01.005

Márton E, Abranches MC, Pais J (2004) Iberia in the Cretaceous: new paleomagnetic results from Portugal. J Geodyn 38:209–221

Matton G, Jébrak M (2009) The Cretaceous Peri-Atlantic Alkaline Pulse (PAAP): deep mantle plume origin or shallow lithospheric break-up? Tectonophysics 469:1–12

Mauffret A, Mougenot D, Miles PR, Malot JA (1989) Cenozoic deformation and Mesozoic abandoned spreading center in the Tagus Abyssal Plain (west of Portugal): results of a multichannel seismic survey. Can J Earth Sci 26(6):1101–1123

Miall AD (1985) Architectural-element analysis: a new method of facies analysis applied to fluvial deposits. Earth-Sci Rev 22:263–308

Miall AD (1996) The geology of fluvial deposits. Sedimentary facies, basin analysis, and petroleum geology. Springer, Heidelberg, p 582

Miranda R, Valadares V, Terrinha P, Mata J, Azevêdo MR, Gaspar M, Kullberg JC, Ribeiro C (2009) Age constrains on the late cretaceous alkaline magmatism on the West Iberian margin. Cretaceous Res 30:575–586

Miskovski JC (sous la direction de) (1987) Géologie de la Préhistoire: méthodes, techniques, applications. Géopré, Edition Association pour l'étude de environnement géologique de la préhistoire, Paris, p 1297

Monegatti P, Raffi S (2001) Taxonomic diversity and stratigraphic distribution of Mediterranean Pliocene bivalves. Palaeog Palaeocl Palaeoec 165:171–193

Moreau MG, Berthou JY, Malod J-A (1997) New paleomagnetic data from the Algarve (Portugal): fast rotation of Ibe' ria between the Hauteverian and the Aptian. Earth Planet Sci Lett 146:689–701

Moreira JD, Lima LP (1987) Prospecção geoeléctrica das argilas de Trás-os-Montes, zona de Palaçoulo (Miranda do Douro). Estudos, Notas e Trabalhos, D.G.G.M 29:71–82

Mougenot D (1989) Geologia da margem portuguesa. Doc. Técn. Instit, Hidrográfico, p 259

Moura D (1998) Litostratigrafia do Neogénico terminal a Plistocénico na Bacia Centro-Algarve, evolução paleoambiental. Ph.D., Univ. Algarve, p 253

Moura D, Boski T (1994) Ludo Formation-a new lithostratigraphic unit in quaternary of central Algarve. GAIA 9:41–47

Moura D, Boski T (1999) Unidadeslitostratigráficas do Pliocénico e Plistocénico no Algarve. Comun Inst Geol e Mineiro 86:85–106

Moya Palomares ME, Azevêdo MT, Rodriguez-Plaza M (2000) Estudio preliminar de los sistemas fluviales cenozoicos de la Cuenca del Guadiana entre Mérida y Badajoz (España). Ciências da Terra 14:223–232

Muñoz JA (1985) Estructura alpina i herciniana a la vora sud de la zona axial del Pirineu oriental. Ph.d. Thesis, Universitat de Barcelona, p 305

Nascimento A (1988) Ostracodos do Miocénico da bacia do Tejo: sistemática, biostratigrafia, paleoecologia, paleogeografia e relações Mediterrâneo-Atlântico. Ph.D., Univ. Nova Lisboa, p 305

Nascimento A (1990) Tentative ostracode biozonation of the Portuguese Neogene. Cour Forsch-Inst Senckenberg 123:181–190

Nascimento A (1993) Application of abundances of Cyprideis group taxa and marine species to the reconstitution of Aquitanian paleoenvironments in Tejo Basin (Portugal). Ostracoda in the earth and life sciencies. A.A. Balkema, Rotterdam, pp 229–239

Ogg J, Lugowski A (2006) Time scale creator 2.0.2. ICS

Pais J (1973) Vegetais fósseis de Ponte de Sor. Bol Soc Geol Portugal 18:123–135

Pais J (1978) Contributions à la Paléontologie du Miocène moyen continental du Bassin du Tage. V-Végétaux de Póvoa de Santarém (note preliminaire). Ciências da Terra 4:103–108

Pais J (1981) Contribuição para o conhecimento da vegetação miocénica da parte ocidental da bacia do Tejo. Ph.D., Univ. Nova Lisboa, p 328

Pais J (1982) O Miocénico do litoral Sul português. Ensaio de síntese. Trabalho complementar obtenção do grau de Doutor, Univ, Nova Lisboa, p 47

Pais J (1986) Évolution de la végétation et du climat pendant le Miocène au Portugal. Ciências da Terra 8:179–191

Pais J (1992) Contributions to the Eocene paleontology and stratigraphy of Beira Alta, Portugal. III - Eocene plant remains from Naia and Sobreda (Beira Alta, Portugal). Ciências da Terra 11:91–108

Pais J (2004) The Neogene of the Lower Tejo Basin (Portugal). Rev Esp Paleontologia 19(2):229–242

Pais J, Legoinha P, Kullberg JC (1991) Novos elementos acerca do Neogénico do Portinho da Arrábida (Serra da Arrábida). III Congresso Nacional de Geologia, Coimbra, p 122

Pais J, Legoinha P, Elderfield H, Sousa L, Estevens M (2000) The Neogene of Algarve (Portugal). Ciências da Terra 14:277–288

Pais J, Silva Lopes C, Legoinha P, Ramalho E, Ferreira J, Ribeiro I, Amado AR, Sousa L, Torres L, Baptista R, Reis RP (2002) The Belverde Borehole (Lower Tagus Basin, Setúbal Peninsula, Portugal). XVIII Jorn. Soc. Esp. Paleontologia, II Cong. ibérico Paleontologia, Interim-Colloquium RCANS. Libro resúmenes, Salamanca, pp 198–199

Pais J, Silva Lopes C, Legoinha P, Ramalho E, Ferreira J, Ribeiro I, Amado AR, Sousa L, Torres L, Baptista R, Reis RP (2003) Sondagem de Belverde (Bacia do Baixo Tejo, Península de Setúbal, Portugal) VI Cong Nacional Geologia. Ciências da Terra n° esp V 13:A99–A102 (CDRom)

Pereira DI (1997) Sedimentologia e estratigrafia do Cenozóico de Trás-os-Montes oriental (NE Portugal). Ph.D., Univ. Minho, p 341

Pereira DI (1998) Enquadramento estratigráfico do Cenozóico de Trás-os-Montes oriental. Comunicações do IGM 84(1):A126–A129 Lisboa

Pereira DI (1999) Terciário de Trás-os-Montes oriental: evolução geomorfológica e sedimentar. Comunicações do IGM 86:213–226 Lisboa

Pereira DI (2006) Depósitos Cenozóicos. In Pereira E (coord), Notícia Explicativa da Folha 2 da Carta Geológica de Portugal na escala 1:200000, Inst. Geol. Mineiro, pp 43–48

Pereira DI, Azevêdo TM (1991) Origem e evolução dos depósitos de cobertura da região de Bragança. Mem. Not. Pub. Mus. Min. Geol. Univ. Coimbra, 112(A): 247–265

Pereira DI, Brilha J (2000) Mineralogia da fracção argilosa da Formação de Vale Álvaro (Bragança, NE Portugal). Ciências da Terra 14:83–88

Pereira DI, Alves MI, Araújo MA, Cunha PP (2000) Estratigrafia e interpretação paleogeográfica do Cenozóico continental do norte de Portugal. Ciências da Terra 14:73–82

Pereira R, Alves TM, Cartwright J (2010) The continent to ocean transition across the SW Iberian margin: The effect of syn-rift geometry on post-Mesozoic compression. Ii Central & North Atlantic Conjugate Margins Conference, Lisbon, pp 228–230

Pimentel NL (1997) O Terciário da Bacia do Sado. Sedimentologia e análise tectono-sedimentar. Ph.D., Univ. Lisboa, p 381

Pimentel NL (1998a) A Formação de Vale do Guizo (Paleogénico) a Sul de Alcácer do Sal. Comun Inst Geol Mineiro 84(1):A149–A152

Pimentel NL (1998b) A Formação de Esbarrondadoiro (Miocénico superior, Bacia do Sado), sedimentologia e paleogeografia. Comun Inst Geol Mineiro 84(1):A152–A156

Pimentel NL (1998c) Tectono-sedimentary evolution of the Sado Basin (Tertiary, southern Portugal). Comun Inst Geol Mineiro 84(1):A145–A148

Pimentel NL (2002) Palaeogene alluvial fan and lacustrine deposits from the Sado Basin (S Portugal). Sed Geol 148(1–2):123–138

Pimentel NL, Azevêdo TM (1990) Terraços fluviais e remobilização das rañas, o exemplo do Rio Sado. Cuatern. y Geomorf 4:119–129

Pimentel NL, Azevêdo TM (1995) Dados para a discussão da génese da Bacia do Tejo-Sado no Paleogénico. Actas IV Cong. Nac. Geologia, pp 897–902

Pinheiro LM, Wilson RCL, Reis RP, Whitmarsh RB, Ribeiro A (1996) The Western Iberia Margin: a geophysical and geological overview. In: Whitmarsh RB, Sawyer DS, Klaus A, Masson D (eds) Proceedings of the ocean drilling program. Scientifics Results 149:3–23

Pisera A, Cachão M, Silva CM (2006) Siliceous sponge spicules from the Miocene Mem Moniz marls (Portugal) and their environmental significance. Riv Itali Paleont Stratig 112(2):287–299

Poag CW, Miller K (1986) In: Vogt PR, Tucholke BE The geology of North América M:547–564

Poças E (2004) Contribuição da palinologia para a caracterização paleoecológica e paleoclimática do Cenozóico a norte do Douro. Msc, Dep. Univ, Minho, p 116

Poças E, Pereira D, Pais J (2003) Análise palinológica preliminar da Formação de Vale Álvaro (Bragança, NE Portugal). Ciências da Terra, n° esp. 5:17–18

Polo MA, Alonso-Gavilán G, Valle MF (1987) Bioestratigraffía y Paleogeografía del Oligoceno-Mioceno del borde SO de la fosa de Ciudad Rodrigo (Salamanca). Stvdia Geologica Salmanticensia 24:229–245

Póvoas L, Brunet-Lecomte P, Chaline J (1995) Présence de Mus spretus fossile dans l'Holocene du portugal. Actas 3ª Reunião Quaternário ibérico, Coimbra, pp 485–489

Rage JC, Augé M (2003) Amphibians and squamate reptiles from the lower Eocene of Silveirinha (Portugal). Ciências da Terra 15:103–116

Ramalhal FJS (1968) Estudo geológico e sedimentológicodos depósitos discordantes dos arredores de Bragança. Publicação do Instituto de Investigação Científica de Angola, Luanda

Ramos A (2008) O Pliocénico e o Plistocénico da plataforma litoral entre o Cabo Mondego e a Nazaré. Ph.D., Univ. Coimbra, p 329

Ramos A, Cunha PP (2004) Facies associations and palaeogeography of the Zanclean-Piacenzian marine incursion in the Mondego cape-Nazaré area (onshore of Central Portugal). Abstracts book of the 23rd International Meeting IAS, Coimbra, p 227

Ramos A, Cunha PP, Gomes A (2009) Os traços geomorfológicos da área envolvente da Figueira da Foz e a evolução da paisagem durante o Pliocénico e o Plistocénico. Public Associação Portuguesa de Geomorfólogos 6:9–16

Rasmussen ES, Lomholt S, Andersen C, Vejbæk OV (1998) Aspects of the structural evolution of the Lusitanian Basin in Portugal and the shelf and slope area offshore Portugal. Tectonophysics, Elsevier 300:199–225

Reis RP (1981) La sédimentation continentale du Crétacé terminal au Miocène sur la Bordadure Occidental du Portugal entre Coimbra et Leiria. Th. 3ème Cycle, Univ. Nancy, p 153

Reis RP (1983) A sedimentologia de depósitos continentais. Dois exemplos do Cretácico Superior—Miocénico de Portugal. Ph.D., Univ. Coimbra, p 404

Reis RP, Cunha PP (1988) Los rellenos terciarios en dos regiones del borde occidental del Macizo Hesperico (Portugal Central). II Congreso Geológico de Espanã, Granada 1:149–152

Reis RP, Cunha PP (1989) Comparacion de los rellenos terciarios en dos regiones del borde occidental del Macizo Hesperico (Portugal Central). Paleogeografia de la Meseta norte durante el Terciário. (Dabrio CJ, Ed.), Stv. Geol. Salman., vol. esp. 5:253–272

Reis RP, Soares AF, Antunes MT (1981) As Areias e Argilas de Silveirinha. Memórias e Notícias, Coimbra 91–92:244–267

Reis RP, Corrochano A, Bernardes C, Cunha PP, Dinis J (1992a) O Meso-Cenozóico da margem atlântica portuguesa. III Congreso Geologico de España, VIII Congreso Latinoamericano de Geologia (Guias de las excursiones geológicas), Salamanca, pp 115–138

Reis RP, Cunha PP, Barbosa BP, Antunes MT, Pais J (1992b) Mainly continental Miocene and Pliocene deposits from Lower Tejo and Mondego Tertiary basins. "Excursion B", Ciências da Terra, n°. Esp 2:37–56

Reis RP, Pais J, Antunes MT (2001) Sedimentação aluvial na região de Lisboa. O "Complexo de Benfica". Geogaceta 29:91–94

Ribeiro A (2002) Soft plate and impact tectonics. Springer Verlag, p 324

Ribeiro A, Antunes MT, Ferreira MP, Rocha R, Soares AF, Zbyszewski G, Moitinho de Almeida F, Carvalho D, Monteiro JH (1979) Introduction à la géologie générale du Portugal. Serviços Geológicos de Portugal, p 114

Ribeiro A, Kullberg MC, Kullberg JC, Manuppela G, Phipps S (1990) A review of Alpine tectonics in Portugal: Foreland detachment in basement and cover rocks. Tectonophysics 184:357–366

Ribeiro A, Kulberg MC, Kullberg JC, Rocha R, Phipps S, Manuppella G (1992) Tectonic inversion of the Lusitanian Basin. 1st Cong Atlantic Neogene Stratig, Lisboa, pp 43–44

Ribeiro A, Cabral J, Baptista R, Matias L (1996) Stress pattern in Portugal mainland and the adjacent Atlantic region, West Iberia. Tectonics 15(2):641–659

Ribeiro O (1942) Notas sobre a evolução morfológica da orla meridional da cordilheira central entre Sobreira Formosa e Fronteira. Bol Soc Geol Portugal I(II):123–145

Ribeiro P, Dias R, Marques FO, Kullberg MC (1996) Estudos de deformação finita na cadeia da Arrábida: primeiros resultados obtidos em amostras de "brecha da Arrábida" colhidas a S do Anticlinal do Formosinho. In: 2.ᵃ Conf. Anual do Grupo de Geologia Estrutural e Tectónica, GGET, pp 24–27

Rodríguez Vidal J, Villalobos M, Jorquera A, Díaz del Olmo F (1988) Geomorfología del sector meridional de la Cuenca del Guadiana. Rev Soc Geol España 1(1–2):157–164

Roest WR, Srivastava SP (1991) Kinematics of the plate boundaries between Eurasia, Iberia and Africa in the North Atlantic from the late Cretaceous to present. Geology 19:613–616

Roman F, Torres A (1907) Le Néogène continental dans la basse vallée du Tage (rive droite). Mém Comiss Serv Geol, Portugal, p 109

Romariz C, Carvalho AMG (1961) Formações margo-glauconíticas do Miocénico superior a Norte do Cabo Espichel. Bol Soc Geol Portugal 14(1):83–94

Romariz C, Correia F, Prates S (1979a) Contributions à la connaissance de l'Algarve (Portugal). III - Un nouveaux facies du Miocène. Bol Mus Lab Min Geol Fac Ciências Lisboa 16(1):265–271

Romariz C, Oliveira M, Almeida C, Baptista R, Cardoso J (1979b) Contributions to the geology of Algarve, Portugal I-The miocene facies of Olhos de Água. Bol Mus Lab Min Geol Fac Ciênc Lisboa 16(1):243–251

Roque C, Terrinha P, Lourenço N, Abreu MP (2009) Morphostructure of the Tore seamount and evidences of recent tectonic activity (West Iberia Margin). 6° Simposio sobre el Margen Ibérico Atlántico MIA09, Oviedo, 1–4

Rosenbaum G, Lister GS, Duboz C (2002) Relative motions of Africa, Iberia and Europe during Alpine orogeny. Tectonophysics 359:117–129

Salgueiro RCM, Martins L (2000) Ocorrência de ouro nos depósitos pliocénicos da região de Cruz de Pau (Seixal). Ciências da Terra 14:203–212

Santisteban JI, Mediavilla R, Martin-Serrano A, Dabrio CJ (1996) The Duero basin: a general overview. In: Friend P, Dabrio J (eds) Tertiary basins of Spain: the stratigraphic record of crustal kinematics. Cambridge University Press, Cambridge, p 400

Santos A (2005) Tafonomia e Paleoicnologia do Neogénico Superior do sector Cacela—Huelva (SE da Ibéria). Ph.D., Univ. Algarve, p 308

Santos A, Boski T (1998) Estudo paleoecológico da Ribeira de Cacela (Miocénico Superior) (Algarve): uma abordagem preliminar. Com do Inst Geol e Mineiro 84(2):A157–A160

Santos A, Boski T, Cachão M, Silva CM, Moura D, Fonseca LC (1998) Jazida fossilífera de Cacela (Parque Natural da Ria Formosa, Algarve): um exemplo de Património Paleontológico a salvaguardar. Com do Inst Geol e Mineiro 84(2):G26–G29

Schermerhorn LJ, Priem HN, Boelrijk NA, Hebeda EH, Verdurmen E, Verschure RH (1978) Age and origin of the Messejana dolerite fault-dike system (Portugal and Spain) in the light of the opening of the North Atlantic Ocean. J Geol 86:299–309

Schettino A, Scotese RC (2002) Global kinematic constraints to the tectonic history of the Mediterranean region and surrounding areas during the Jurassic and Cretaceous. In: Rosenbaum G, Lister GS (eds) Reconstruction of the evolution of the Alpine-Himalayan orogen. J Virt Expl 7:147–166

Schettino A, Turco E (2011) Tectonic history of the western Tethys since the Late Triassic. Geol Soc America Bull 123(1–2):89–105

Schott JJ, Montigny R, Thuizat R (1981) Paleomagnetism and potassium–argon age of the Messejana dyke (Portugal and Spain): angular limitation to the rotation of the Iberian Peninsula since the middle Jurassic. Earth Planet Sci Lett 53:457–470

Sen S, Antunes MT, Pais J, Legoinha P (1992) Bio and magnetostratigraphy of two Lower Miocene sections, Tejo basin (Portugal). Ciências da Terra 11:173–184

Sequeira A, Cunha PP, Sousa MB (1997) A reactivação de falhas, no intenso contexto compressivo desde meados do Tortoniano, na região de Espinhal-Coja-Caramulo (Portugal Central). Com Inst Geol Mineiro 83:95–126

Sharpe D (1834) On the strata in the immediate neighbourhood of Lisbon and Oporto. Proc Geol Soc London 1(1826–1833):394–396

Sharpe D (1841) On the geology of the neighbourhood of Lisbon. Transactions of the Geol. Soc. London, VI, sections 1, 9, 11, 12 and 13

Sierro FJ (1984) Foraminíferos planctónicos y bioestratigrafía del Mioceno superior -Plioceno del borde occidental de la cuenca de Guadalquivir (S.O. de España). Ph.D., Univ. Salamanca, p 391

Sierro FJ (1985) The replacement of the Globorotalia menardii group by the Globorotalia miotumida group: an aid to recognizing the Tortonian-Messinian boundary in the Mediterranean and the adjacent Atlantic. Mar Micropaleontol 9:525–535

Sierro FJ, Flores JA (1992) Evolución de las fosas bética y rifeña y la comunicación Atlántico-Mediterráneo durante el Mioceno. Simposios III Congr. Geol. España, Salamanca 2:563–567

Sierro FJ, Flores JA, Bárcena MA, Civis J, González Delgado JA (1989) Afloramientos de aguas profundas atlánticas en el estrecho Norbético: implicación en la dinámica Atlántico-Mediterráneo durante el Mioceno. Resum. V Jorn. Paleont.Valencia, pp 147–148

Sierro FJ, Flores JA, Civis J, González Delgado JA, Francés G (1993) Late Miocene globorotaliid event-stratigraphy and biogeography in the NE-Atlanic and Mediterranean. Mar Micropaleontol 21:143–168

Sierro FJ, González Delgado JA, Dabrio CJ, Flores JA, Civis J (1996) Late Neogene depositional sequences in the foreland basin of Guadalquivir (SW Spain). In: Friend PF, Dabrio CJ (eds) Tertiary basins of Spin. Cambridge University Press, Cambridge, pp 339–345

Silva MJBL (1988) Hidrogeologia do Miocénico do Algarve. Ph.D., Univ. Lisboa, p 375

Silva CM (2001) Gastrópodes Pliocénicos Marinhos de Portugal. Sistemática, Paleoecologia, Paleobiologia, Paleobiogeografia. Ph.D., Univ. Lisboa, p 747

Silva CM, Landau B, Domènech R, Martinell J (2010) Pliocene Atlantic molluscan assemblages from the Mondego Basin (Portugal): age and palaeoceanographic implications. Palaeogeogr Palaeoclimatol Palaeoecol 285:248–254

Soares AF, Reis RP (1982) Esboço de enquadramento cronostratigráfico das unidades líticas pós-jurássicas da Orla Meso-Cenozóica Ocidental entre os paralelos de Pombal e Aveiro. Memórias e Notícias, Coimbra 93:77–91

Soares AF, Lapa M, Marques J (1986) Contribuição para o conhecimento da litologia das unidades rneso-cenozóicas da Bacia Lusitaniana a Norte do "acidente" da Nazaré (Sub-zona Setentrional). Memórias e Notícias, Coimbra 102:23–41

Soulié-Marsh I (1978) Contributions à la Paléontologie du Miocène moyen continental du Bassin du Tage. IV - Charophytes - Póvoa de Santarém, Pero Filho e Tremês. Ciências da Terra 4:91–102

Sousa PFL (1917) Sur les eruptions du littoral de l'Algarve (Portugal). CR Acad Sci Paris 165:674–675

Sousa PFL (1922) Sur les roches eruptives de la bordure mesozo et cainozo de l'Algarve et leur âge géologique. CR Acad Sci Paris 175:822–824

Srivastava SP, Roest WR, Kovacs LC, Oakey G, Levesque S, Verhoef J, Macnab R (1990) Motion of Iberia since the late Jurassic: results from detailed aeromagnetic measurements in the Newfoundland Basin. Tectonophysics 184:229–260

Suc JP (1984) Origin and evolution of the meditteranean vegetation and climate in Europe. Nature 307:429–432

Suc JP, Bertini A, Combourieu-Nebout N, Diniz F, Leroy S, Russo-Ermolli E, Zheng Z, Bessais E, Ferrier J (1995) Structure of West Mediterranean vegetation and climate since 5, 3 ma. Acta zool. Cracov 38:3–16

Swift DJP, Thorne JA (1991) Sedimentation on continental margins, I: a general model for shelf sedimentation. In: Swift DJP, Oertel GF, Tillman RW, Thorne JA (eds) Shelf sands and sandstone bodies. Geometry, facies and sequence stratigraphy. IAS Special Publications, vol 14. Blackwell Scientific Publications, Oxford, pp 3–31

Teixeira C (1952) Flora fóssil do Miocénico de Esbarrondadoiro, Odivelas. Comun Serv Geol Port 33:93–97

Teixeira C, Pais J (1976) Introdução à paleobotânica. As grandes fases da evolução dos vegetais. Ed. autores, p 210

Tejero R, Garzón Heydt G, Babín Vich R, Fernández García P (2010) Long-Term evolving "Tectonic" landscapes within intra-plate domains. The Iberian Peninsula. Horizons in Earth Science Research, 2. Cap 4. Nova Publishers Inc, USA, pp 103–124

Terrinha P (1998) Structural geology and tectonic evolution of the Algarve Basin, South Portugal. Ph.D., Imperial College, London, p 430

Terrinha P, Kullberg JC, Kullberg MC, Moita C, Ribeiro A (1996) Thin skinned and thick skinned sub-basin development, bidimensional extension and self-indentation in the Lusitanian Basin, West Portugal. 2ª Conf Nac. GGET, Soc. Geol. Portugal, pp 17–20

Terrinha P, Matias L, Vicente J, Duarte J, Luís J, Pinheiro L, Lourenço N, Diez S, Rosas F, Magalhães V, Valadares V, Zitellini N, Roque C, Mendes Víctor L, MATESPRO Team (2009) Morphotectonics and strain partitioning at the Iberia–Africa plate boundary from multibeam and seismic reflection data. Marine Geol 267:156–174

Terrinha P, Ribeiro C, Kullberg JC, Lopes C, Rocha R, Ribeiro A (2002) Compressive episodes and faunal isolation during rifting, southwest Iberia. J Geol 110:101–113

Terrinha P, Rocha R, Rey J, Cachão M, Moura D, Roque C, Martins L, Valadares V, Cabral J, Azevedo MR, Barbero L, Clavijo E, Dias RP, Gafeira J, Matias H, Matias L, Madeira J, Marques da Silva C, Munhá J, Rebelo L, Ribeiro C, Vicente J, Youbi N (2006) A Bacia do Algarve: estratigrafia, Paleogeografia e tectónica. In: Dias R, Araújo A, Terrinha P, Kullberg JC (eds) Geologia de Portugal no contexto da Ibéria. Univ Évora, Évora, pp 247–316

Terrinha P, Rocha R, Rey J, Cachão M, Moura D, Roque C, Martins L, Valadares V, Cabral J, Azevedo MR, Barbero L, Clavijo E, Dias RP, Matias H, Madeira J, Marques da Silva C, Munhá J, Rebelo L, Ribeiro C, Vicente J, Noiva J, Mata J, Youbi N, Bensalah MK (2011) A Bacia do Algarve: Estratigrafia, Paleogeografia e Tectónica. In: Geologia de Portugal no contexto da Ibéria, Escolar Editora (in press)

Truc G (1977) Contributions à la Paléontologie du Miocène moyen continental du Bassin du Tage. I–Quelques mollusques - Pero Filho, Póvoa de Santarém, Sítio do Mirante. Ciências da Terra 3:121–127

Vallin S (1965) Sur une cupressaceae fossile du Portugal. Bol Soc Geol Portugal 26:111–124

Vegas R (1989) Kinematic constrains and Alpine-Present tectonics of Iberia. Annales Geophysicae, sp. issue. Gauthier-Villars, Paris, p 26

Vegas R (2005) Deformación alpina de macizos antiguos: el caso del Macizo Ibérico (Hespérico). Bol Real Soc Esp Hist Nat sec geológica 100(1–4):39–54

Vegas R (2006) Modelo tectónico de formación de los relieves montañosos y las cuencas de sedimentación terciarias del interior de la Península Ibérica. Bol R SocEsp Hist Nat (Sec Geol) 101(1–4):31–40

Veiga-Pires C, Moura D, Rodrigues B, Machado N, Campo L, Simonetti A (2007) Provenance of Quaternary sands in the Algarve (Portugal) revealed by U-Pb ages of detrital zircon. In: Nichols G, Williams E, Paola C (eds) Sedimentary Processes, Environments and basins: A tribute to Peter Friend Special publication number 38 of the International Association of Sedimentologists. Blackwell, Oxford, pp 327–340

Verhoef J, Srivastava SP (1989) Correlation of sedimentarybasins across the North Atlantic as obtained from gravity andmagnetic data, and its relation to the early evolution of the North

Atlantic. In: Tankard AJ, Balkwill HR (eds) Extensional Tectonics and Stratigraphy of the North AtlanticMargins. Am Assoc Pet Geol Mem 46:131–148

Vieira MC (2009) Palinologia do Pliocénico da Orla Ocidental Norte e Centro de Portugal: Contributo para a compreensão da cronostratigrafia e da evolução paleoambiental. Ph.D., Univ. Minho, p 389

Villalobos M, Jorquera A (1998) El Terciario continental y Cuaternario del sector meridional de la Cuenca del Guadiana. Publicaciones del Museo Geológico de Extremadura, pp 33–44

Villalobos M, Jorquera A, Apalategui I (1988) Hoja núm. 802 (La Albuera), 1:50.000 segunda serie. IGME

Villamor P (2002) Cinemática Terciaria y Cuaternaria de la Falla de Alentejo-Plasencia y su influencia en la peligrosidad sísmica de la Península Ibérica. Ph.D., Univ. Complutense Madrid, p 343

Wilson RCL, Hiscott RN, Willis MG, Gradstein FM (1989) The Lusitanian Basin of West Central Portugal: Mesozoic and Tertiary Tectonic, Stratigraphic, and Subsidence History. In Tankard AJ, Balkwill H (eds) Extensional tectonics and stratigraphy of the North Atlantic margins. AAPG Memoir 46:341–361

Witt WG (1977) Stratigraphy of the Lusitanian Basin 61 Unpublished Shell Prospex Portuguesa Report

Zagwijn WH (1960) Aspects of the Pliocene and early Pleistocene vegetation in the Netherlands. Meded Geol Schicht ser C 3(5):78

Zbyszewski G (1941) Les problèmes du Néogène continental de la basse vallée du Tage (rive droite). X Cong. Asociación Esp. Progreso Ciencias, Zaragosa, p 25

Zbyszewski G (1943) Élements pour servir a l'étude du Pliocène marin du Sud du Tage: la faune des couches supérieur d'Alfeite. Com Serv Geol Portugal 24:125–156

Zbyszewski G (1948) O Miocénico marinho de Bensafrim (Algarve). Bol Soc Geol Portugal Porto 7:55–56

Zbyszewski G (1950) Les restes d'hippopotame et des cerfs d'Algoz. Com Serv Geol Portugal Lisboa 31:413–422

Zbyszewski G (1954) L'Aquitanien supérieur de Lisbonne et du Ribatejo. Com Serv Geol Portugal 35:99–154

Zbyszewski G (1957) Le Burdigalien de Lisbonne. Com Serv Geol Portugal 38:91–215

Zbyszewski G (1962) Considérations sur la position stratigraphique de l'Aquitanien portugais. Com Serv Geol Portugal 46:297–316

Zbyszewski G (1963) Carta geológica dos arredores de Lisboa na escala de 1/50000. Notícia Explicativa da folha 4 (Lisboa). Serv Geol Portugal, p 93

Zbyszewski G (1964) Les rapports entre les milieux miocènes marins et continentaux au Portugal. Instituto Lucas Mallada, CSIC (España), Cursillos y Conferencias 9:103–108

Zbyszewski G (1967) Contributions à l'étude du Miocène de la serra da Arrábida. Com Serv Geol Portugal Lisboa 51:37–148

Zbyszewski G, Almeida FM (1960) Carta Geológica de Portugal na escala 1/50000. Notícia explicativa da folha 26-D (Caldas da Rainha). Serv Geol Portugal, p 56

Zbyszewski G, Ferreira OV (1967) Découverte de vértébrés fossiles dans le Miocène de la région de Leiria. Com serv Geol Portugal 52:5–10

Zbyszewski G, Teixeira C (1949) Le niveaux quaternaire marin de 5–8 mètres au Portugal. Bol Soc Geol Portugal 8(1/2):1–6

Zbyszewski G, Ferreira OV, Manupella G, Assunção CT (1965) Carta geológica de Portugal na escala 1/50000. Notícia explicativa da folha 38-B (Setúbal). Serv. Geol. Portugal, Lisboa. p 134

Zitellini N, Gràcia E, Matias L, Terrinha P, Abreu MA, DeAlteriis G, Henriet J-P, Dañobeitia JJ, Masson DG, Mulder T, Ramella R, Somoza L, Diez S (2009) The quest for the Africa-Eurasia plate boundary west of the Strait of Gibraltar. Earth and Planetary Science Letters, p 28

Appendix

Localities

Locality	Municipality	Geological unit	Latitude	Longitude W (-)	Altitude (m)
Abrantes	Abrantes	Paleozoic	39° 27′ 52″	8° 11′ 53″	185
Aire mountain	Torres Novas	Mesozoic	39° 32′ 00″	8° 38′ 00″	650
Albandeira	Armação de Pera	Lagos-Portimão Fm.	37° 05′ 26″	8° 24′ 01″	5
Albardeira	Lagos	Lagos-Portimão Fm.	37° 07′ 07″	8° 39′ 50″	25
Albufeira	Albufeira	Lagos-Portimão Fm.	37° 05′ 13″	8° 15′ 11″	10
Albufeira lagoon	Sesimbra	Holocene	38° 30′ 42″	9° 10′ 28″	0
Alcácer do Sal	Alcácer do Sal	Middle Miocene	38° 22′ 28″	8° 03′ 53″	15
Alcanede	Santarém	Monsanto Fm.	39° 24′ 50″	8° 49′ 24″	75
Alcanena	Alcanena	Almoster-Santarém Fm.	39° 27′ 35″	8° 40′ 02″	75
Alcobaça	Alcobaça	Upper Jurassic	39° 32′ 38″	8° 58′ 41″	55
Alcoentre	Azambuja	Alcoentre Fm.	39° 12′ 32″	8° 58′ 34″	85
Alfundão	Ferreira do Alentejo	Esbarrondoiro Fm.	38° 7′ 10″	8° 03′ 44″	110
Algoz	Silves	Pleistocene	37° 09′ 49″	8° 18′ 17″	50
Aljezur (Furna Amarela)	Aljezur	Midde Miocene	37° 21′ 42″	8° 46′ 52″	50
Aljezur (Igreja Nova)	Aljezur	Lower Miocene	37° 19′ 16″	8° 47′ 48″	25
Aljezur (Gimnosdesportivo)	Aljezur	Lower Miocene	37° 19′ 35″	8° 47′ 38″	25

(continued)

J. Pais et al., *The Paleogene and Neogene of Western Iberia (Portugal)*,
SpringerBriefs in Earth Sciences, DOI: 10.1007/978-3-642-22401-0,
© João Pais 2012

(continued)

Locality	Municipality	Geological unit	Latitude	Longitude W (-)	Altitude (m)
Almada	Almada	Lower Miocene	38° 40′ 49″	9° 09′ 30′	50
Almeirim	Almeirim	Ulme Fm	39° 12′ 35″	8° 37′ 43″	15
Almeirim Mountain	Almeirim	Almeirim Fm.	39° 08′ 39″	8° 35′ 37″	150
Almoster	Santarém	Almoster-Santarém Fm.	39° 14′ 30″	8° 47′ 35″	30
Alto de S. João	Lisboa	Div. Va2	38° 43′ 43″	9° 07′ 22″	80
Amareleja	Moura	Moura Basin deposits	38° 12′ 16″	7° 12′ 36″	220
Amor	Leiria	Amor Fm.	39° 48′ 11″	8° 51′ 59′	35
Archino	Azambuja	Alcoentre Fm.	39° 06′ 17″	8° 56′ 24″	40
Arganil	Arganil	Paleozoic	40° 13′ 10″	8° 03′ 14″	195
Armação de Pêra	Silves	Lagos-Portimão Fm.	37° 06′ 04″	8° 21′ 43″	5
Arrábida	Setúbal	Jurassic, Paleogene and Neogene	38° 28′ 24″	8° 59′ 33″	490
Arrifana	Macedo de Cavaleiros	Aveleda Fm.	41° 34′ 34″	6° 55′ 40″	650
Arrifão	Albufeira (Faro)	Lagos-Portimão Fm.	37° 04′ 44″	8° 16′ 01″	0
Arroça	Arganil	Serra de Sacões Group	40° 12′ 44″	8° 10′ 20″	370
Atalaia	Bragança (norte)	Bragança Fm, Atalaia Mb.	41° 50′ 09″	6° 44′ 24″	655
Atenor	Miranda do Douro	Bragança Fm.	41° 25′ 44″	6° 28′ 33″	674
Auramar Hotel beach	Albufeira (Faro)	Lagos-Portimão Fm / Fine sands	37° 05′ 00″	8° 13′ 52″	0
Av. do Uruguai	Lisboa	Divisão II	38° 45′ 13″	9° 11′ 42″	77
Aveleda	Aveleda, Bragança	Aveleda Fm.	41° 53′ 40″	6° 42′ 35″	825
Ávila	Ávila (Spain)	Paleozoic	40° 39′ 23″	4° 42′ 01″	1105
Azambuja	Azambuja	Alcoentre Fm.	39° 04′ 07″	8° 52′ 09″	20
Azambujeira	Rio Maior	Aalcoentre Fm.	39° 16′ 14″	8° 47′ 35″	75
Azeitão	Sesimbra	Santa Marta Fm.	38° 31′ 06″	9° 00′ 44″	110
Azibeiro (falha)	Macedo de Cavaleiros	Aveleda Fm.	41° 36′ 27″	6° 54′ 05″	649
Badajoz	Badajoz (Spain)	Guadiana Basin	38° 52′ 43″	6° 58′ 13″	180
Barca D'Alva	Figueira de Castelo Rodrigo	Paleozoic	41° 01′ 38″	6° 56′ 26″	145
Barosa	Leiria	Barracão Group	39° 45′ 24″	8° 51′ 06″	75
Barracão	Leiria	Barracão Group	39° 49′ 32″	8° 43′ 20′	175
Barreiro	Barreiro	Pliocene	38° 39′ 51″	9° 04′ 36″	4

(continued)

(continued)

Locality	Municipality	Geological unit	Latitude	Longitude W (-)	Altitude (m)
Batalha	Leiria	Mesozoic	39° 39′ 28″	8° 49′ 27″	73
Batocas	Nave de Haver, Almeida	Vilariça arkoses	40° 28′ 26″	6° 50′ 47″	799
Belverde	Seixal	Belverde conglomerates	38° 35′ 55′	9° 08′ 22″	49
Belverde	Alcácer do Sal	Paleozoic socle	38° 19′ 37″	8° 31′ 31″	73
Benfica	Lisboa	Benfica Fm.	38° 44′ 57″	9° 12′ 04″	75
Bensafrim	Lagos	Lagos-Portimão Fm., Paleozoic	37° 09′ 28″	8° 44′ 05″	30
Besteiros	Tondela	Coja Fm	40° 33′ 28″	8° 08′ 10″	275
Braço de Prata	Lisboa	Div. VIIa, VIIb	38° 45′ 09″	9° 05′ 59″	20
Bragança	Bragança	Lagos-Portimão Fm.	41° 48′ 21″	6° 39′ 50″	660
Cabanas	Palmela	Marco Furado Fm	38° 33′ 22″	8° 58′ 08″	100
Cabanas de Cima	Vilariça, Torre de Moncorvo	Vilariça arkoses	41° 12′ 46″	7° 06′ 31″	135
Cabeço do Infante	Castelo Branco	Cabeço do Infante Fm.	39° 50′ 25″	7° 39′ 46″	350
Cabo Mondego	Figueira da Foz	Mesozoic	40° 11′ 20″	8° 54′ 07″	143
Cabo Ruivo	Lisboa	Div. VIIb	38° 45′ 21″	9° 06′ 08″	35
Cacela	Vila Real de Santo António	Cacela Fm.	37° 09′ 26″	7° 32′ 54″	24
Cacilhas	Almada	Burdigalian	38° 41′ 11″	9° 08′ 55″	15
Caldas da Rainha	Caldas da Rainha	Pliocene	39° 24′ 28″	9° 08′ 02″	55
Campelo	Góis	Campelo Fm.	40° 12′ 03″	8° 10′ 15″	300
Campina de Faro	Faro	Fine sands and sandstones	37° 02′ 58″	7° 55′ 44″	20
Campo Grande	Lisboa	Burdigalian	38° 44′ 60″	9° 09′ 25″	93
Campo Maior	Campo Maior	Paleozoic	39° 00′ 36″	7° 04′ 06″	270
Canavial beach	Lagos	Lagos-Portimão Fm.	37° 05′ 29″	8°19′ 48″	0
Candeeiros Mountain	Rio Maior	Mesozoic	39° 26′ 00″	8° 55′ 00″	435
Capuchos	Almada	Tortonian	38° 38′ 36″	9° 13′ 23″	58
Caramulo Mountain	Aveiro	Paleozoic socle	40° 34′ 19″	8° 10′ 12″	950
Carcavelos	Cascais	Lower Miocene	38° 40′ 41″	9° 19′ 51″	3
Carnide	Leiria	Carnide Fm.	39° 52′ 33″	8° 43′ 52″	100
Carnide (Lisboa)	Lisboa	Lower Miocene	38° 45′ 39″	9° 11′ 01″	100
Carviçais	Torre de Moncorvo	Aveleda Fm.	41° 11′ 05″	6° 51′ 56″	652

(continued)

(continued)

Locality	Municipality	Geological unit	Latitude	Longitude W (-)	Altitude (m)
Carvoeiro	Lagoa	Lagos-Portimão Fm.	37° 05′ 46″	8° 28′ 18″	25
Casais da Formiga	Azambuja	Alcoentre Fm.	39° 04′ 39″	8° 54′ 04″	95
Casal das Chitas	Lisboa	Div. Vb, Vc	38° 46′ 57″	9° 08′ 32″	130
Casal Vistoso	Lisboa	Divs. IVb, Va1	38° 44′ 33″	9° 07′ 49″	95
Castelãos	Macedo de Cavaleiros	Aveleda Fm.	41° 31′ 31″	6° 55′ 31″	562
Castelo beach	Albufeira	Lagos-portimão Fm.	37° 04′ 24″	8° 17′ 55″	5
Castelo Branco	Castelo Branco	Paleozoic	39° 49′ 26″	7° 29′ 31″	400
Castro	Bragança	Bragança Fm., Castro Mb.	41° 48′ 16″	6° 47′ 20	719
Castro Roupal	Macedo de Cavaleiros	Bragança Fm.	41° 32′ 21″	6° 48′ 34″	594
Chão da Anixa	Setúbal	Burdigalian	38° 29′ 00″	8° 58′ 23″	50
Charneca do Lumiar	Lisboa	Divs. Vb-c	38° 47′ 08″	9° 08′ 46″	125
Chaves	Chaves	Paleozoic and Pleistocene	41° 44′ 29″	7° 27′ 36″	350
Chelas	Lisboa	Middle Miocene	38° 45′ 04″	9° 06′ 56″	70
Coelha beach	Albufeira	Lagos-Portimão Fm.	37 04′ 25″	8 17′ 49″	5
Coelhos beach	Setúbal	Lower Miocene	38° 29′ 01″	8° 58′ 09″	5
Coimbra	Coimbra	Mesozoic	40° 12′ 41″	8° 25′ 45″	60
Coja	Arganil	Coja Fm.	40° 16′ 02″	7° 59′ 10″	185
Costa de Caparica	Almada	Middle and Upper Miocene	38° 38′ 45″	9° 14′ 00″	5
Cotovia	Sesimbra	Lower Miocene	38° 28′ 43″	9 06′ 15″	125
Crete	Crete (Greece)	Miocene	35° 14′ 24″	E(+)24° 48′ 33″	1580
Cristo Rei	Almada	Divs. II, III, IVa, IVb, Va1, Va2, Va3	38° 40′ 43″	9° 10′ 17″	133
Cruz de Pau	Seixal	Santa Marta Fm.	38° 37′ 18″	9° 07′ 16″	30
Eixes	Mirandela	Mirandela Fm.	41° 30′ 14″	7° 12′ 10″	250
Elvas	Elvas	Paleozoic / Guadiana Basin	38° 52′ 52″	7° 09′ 50″	290
Espichel Cape	Sesimbra	Mesozoic	38° 24′ 49″	9° 13′ 10″	100
Estrela mountain	Covilhã	Paleozoic socle	40° 21′ 09″	7° 34′ 54″	1990
Fábrica	Vila Real de Santo António	Cacela Fm.	37° 09′ 21″	7° 33′ 09″	24
Falagueira	Nisa	Falagueira Fm.	38° 31′ 27″	7° 44′ 59″	300

(continued)

(continued)

Locality	Municipality	Geological unit	Latitude	Longitude W (-)	Altitude (m)
Falésia beach	Loulé	Ludo Fm.	37° 05′ 10″	8° 10′ 04″	5
Faro	Faro	Upper Miocene and Pliocene	37° 00′ 54″	7° 56′ 2,0″	10
Feligueira Grande	Alcobaça	Bom Sucesso Fm.	39° 40′ 44″	9° 03′ 36″	10
Figueira	Portimão	Igneous dike	37° 09′ 35″	8° 35′ 56″	25
Figueira Brava cave	Sesimbra	Lower Miocene	38° 28′ 13″	8° 59′ 11″	4
Figueiras	Chamusca	Ulme Fm., Almeirim Fm..	39° 18′ 18″	8° 26 24″	140
Flamenga	Setúbal	Guarda Mor Cong.	38° 32′ 20″	8° 54′ 52″	50
Folques	Arganil	Serra de Sacões Group	40° 13′ 26″	8° 00′ 32′	240
Fonte da Telha	Almada	Santa Marta Fm.	38° 34′ 27″	9° 11′ 44″	15
Forte da Baralha	Sesimbra	Mesozoic	38 24′ 47″	9° 11′ 25″	40
Foz da Fonte	Sesimbra	Burdigalian	38° 27′ 12″	9° 11′ 57″	5
Foz do Rêgo	Almada	Tortonian	38° 37′ 30″	9° 12′ 42″	30
Galé beach	Albufeira	Lagos-Portimão Fm.	37° 04′ 50″	8° 18′ 58″	7
Galvana	Faro	Galvana conglomerates	37° 02′ 56″	7° 54′ 47″	25
Gambelas	Faro	Gambelas Member	37° 02′ 32″	7° 58′ 506′	23
Gavião	Gavião	Ulme Fm.	39° 27′ 49″	7° 55′ 33″	215
Goldra	Loulé	Karst infill	37° 07′ 55″	8° 00′ 17″	200
Grilos	Lisboa	Div. VIb	38° 43′ 46″	9° 06′ 30″	15
Guarda	Guarda	Paleozoic	40° 32′ 14″	7° 16′ 04″	1015
Guia	Albufeira	Paleogene?	37 07′ 41″	8 18′ 03″	50
Horta das Tripas	Lisboa	Div. I	38° 43′ 51″	9° 08′ 42″	80
Hotel Auramar beach	Albufeira	Lagos-Portimão Fm., Fine sands	37° 05′ 01″	8° 13′ 51″	5
Juromenha	Alandroal	Guadiana basin deposits	38° 44′ 28″	7° 14′ 23″	195
Lagos	Lagos	Lagos-Portimão Fm.	37° 05′ 57″	8° 40′ 23″	25
Lapa de Santa Margarida	Sesimbra	Lower Miocene	38° 28′ 12″	8° 59′ 12″	4
Laranjeiro	Almada	Santa Marta Fm.	38° 39′ 24″	9° 08′ 59″	40
Lardosa	Castelo Branco	planation surface on granites	39° 59′ 22″	7° 26′ 38″	400
Leiria	Leiria	Mesozoic	39° 44′ 38″	8° 48′ 27″	50
Limãos	Macedo de Cavaleiros	Vale Álvaro Fm.	41° 31′ 44″	6° 49′ 33″	603

(continued)

(continued)

Locality	Municipality	Geological unit	Latitude	Longitude W (-)	Altitude (m)
Lisboa	Lisboa	Benfica Fm. and Miocene deposits	38° 43′ 34″	9° 08′ 03″	50
Lisboa (Alto dos Moinhos)	Lisboa	Lower Miocene	38° 45′ 02″	9° 10′ 47″	88
Lisboa (Av. Padre Cruz)	Lisboa	Div. IVa	38° 45′ 59″	9° 09′ 56″	110
Lisboa (Av. Uruguai)	Lisboa	Div. II	38° 45′ 14″	9° 11′ 42″	75
Lisboa (Chelas)	Lisboa	Divs. Vb, Vc	38° 45′ 04″	9° 06′ 56″	70
Lisboa (Parque Eduardo VII)	Lisboa	Div. I	38° 43′ 57″	9° 09′ 11″	90
Lisboa (Ralis)	Lisboa	Div. VIc	38° 46′ 56″	9° 07′ 06″	65
Lisboa (Univ. Católica)	Lisboa	Div. II	38° 44′ 57″	9° 09′ 57″	85
Lobão da Serra	Tondela	Lobão Fm.	40° 31′ 41″	8° 02′ 28″	300
Longroiva (Termas fault)	Longroiva, Meda	Vilariça arkoses	40° 57′ 48″	7° 12′ 03″	315
Longroiva, EN 102	Longroiva, Meda	Vilariça arkoses	40° 57′ 38″	7° 11′ 49″	279
Loures	Loures	Paleogene	38° 49′ 49″	9° 10′ 02″	20
Lousã	Coimbra	Paleozoic basement	40° 07′ 03″	8° 14′ 35″	170
Lousã mountain	Lousã	Paleozoic basement	40° 04′ 00″	8° 13′ 00″	750
Lucefeci	Évora	Pliocene	38° 38′ 08″	7° 25′ 26″	175
Ludo	Faro	Ludo Member	37° 01′ 49″	7° 59′ 48″	7
Lumiar	Lisboa	Burdigalian and Langhian	38° 45′ 55″	9° 09′ 30″	100
Luz de Tavira	Tavira	Morgadinho deposits	37° 05′ 29″	7° 42′ 17″	30
Macedo de Cavaleiros	Macedo de Cavaleiros	Bragança Fm.	41° 32′ 11″	6° 57′ 22″	550
Madrid	Madrid, Spain	Cenozoic Height Tejo Basin	40° 25′ 00″	3° 42′ 01″	650
Mafra	Mafra	Mesozoic	38° 56′ 13″	9° 19′ 35″	235
Magarefe Mountain	Castelo Branco	Falagueira Fm.	39° 53′ 00″	7° 42′ 00″	355
Malpica	Castelo Branco	Paleozoic	39° 40′ 38″	7° 23′ 44″	320
Manteigas	Manteigas	Paleozoic	40° 24′ 02″	7° 32′ 23″	750
Marco Furado	Palmela	Marco Furado Fm.	38° 34′ 45″	9° 01′ 34″	50
Maria Luísa beach	Albufeira	Lagos-Portimão Fm.	37° 05′ 20″	8° 12′ 01″	5
Marmelar	Vidigueira	Moura Basin deposits	38° 10′ 15″	7° 39′ 14″	115

(continued)

(continued)

Locality	Municipality	Geological unit	Latitude	Longitude W (-)	Altitude (m)
Mêda	Mêda	Vilariça Fm.	40° 57′ 47″	7° 15′ 47″	730
Meia Praia	Lagos	Cacela Fm.	37° 07′ 03″	8° 38′ 51″	5
Mem Moniz	Albufeira	Mem Moniz spongoliths	37° 09′ 53″	8° 13′ 09″	75
Mérida	Mérida (Spain)	Paleozoic	38° 54′ 57″	6° 20′ 44″	205
Messejana	Aljustrel	Paleozoic	37° 50′ 09″	8° 14′ 53″	195
Miranda do Douro	Miranda do Douro	Paleozoic	41° 29′ 45″	6° 16′ 26″	650
Mirandela	Mirandela	Mirandela Fm.	41° 29′24″	7° 10′ 39″	230
Monchique Mountain	Monchique	Igneous sub-volcanic	37° 18′ 53″	8° 35′ 44″	890
Monforte da Beira	Castelo Branco	Cabeço do Infante Fm.	39° 44′ 24″	7° 17′ 45″	300
Monfortinho	Idanha a Nova	Monfortinho Fm., Torre Form.	39° 59′ 50″	6° 52′ 50″	275
Monsanto	Idanha a Nova	granite inselberge	40° 02′ 00″	7° 06′ 44″	668
Monte de Caparica	Almada	Div. VIa-VIIa	38° 39′ 46″	9° 11′ 48″	95
Montenegro	Faro	Montenegro sands	37° 01′ 54″	7° 57′ 40″	25
Montejunto	Cadaval	Jurassic	39° 10′ 25″	9° 02′ 56″	650
Montesinho	Bragança	Paleozoic	41° 47′ 47″	6° 32′ 31″	640
Montesinho Mountain	Bragança	Paleozoic	42° 02′ 25″	6° 51′ 39″	1720
Montijo	Montijo	Santa Marta Fm.	38° 42′ 21″	8° 58′ 29″	5
Montoito	Redondo	Pliocene	38° 30′ 31″	7° 35′ 48″	230
Moradal	Oleiros	Paleozoic (quartzite inselberge)	39° 59′ 00″	7° 47′ 00″	830
Morgadinho	Tavira	Morgadinho deposits	37° 05′ 50″	7° 42′ 43″	33
Morouços	Coimbra	St. Quitéria Fm.	40° 10′ 53″	8° 27′ 07″	205
Mortágua	Mortágua	Viseu	40° 19′ 36″	8° 15′ 01″	135
Moura	Moura	Moura Basin deposits	38° 08′ 39″	7° 26′ 58″	140
Murracha	Idanha a Nova	Murracha Group	39° 57′ 00″	7° 02′ 00″	400
Murrachinha	Idanha a Nova	Murracha Group	39° 58′ 52″	6° 58′ 58″	505
Naia	Tondela	Coja Fm.	40° 31′ 14″	8° 04′ 31″	300
Nave de Haver	Almeida	Paleogene	40° 31′ 18″	6° 50′ 03″	820
Nazaré	Leiria	Cretaceous	39° 36′ 32″	9° 02′ 06″	50
Nisa	Nisa	planation surface on granite	39° 31′ 05″	7° 39′ 12″	300

(continued)

(continued)

Locality	Municipality	Geological unit	Latitude	Longitude W (-)	Altitude (m)
Nogueira Mountain	Bragança	Paleozoic	41° 43' 02"	6° 51' 17"	1300
Odemira	Odemira	Esbarrondadoiro Fm ?	37° 36' 29"	8° 36' 06"	60
Olhão	Olhão	Pleistocene - Holocene	37° 01' 36"	7° 50' 26"	5
Olhos de Água	Albufeira	Lagos-Portimão Fm.	37° 05' 25"	8° 11' 22"	10
Olival da Susana	Lisboa	Divs. Vb, Vc	38° 46' 52"	9° 08' 13"	100
Oriola	Portel	Moura Basin deposits	38° 19' 10"	7° 51' 47"	200
Ourém	Ourém	Serra de Sacões Group	39° 38' 35"	8° 35' 29"	265
Óvoa	Santa Comba Dão	Lobão Fm.	40° 22' 50"	8° 07' 52"	200
Palença	Almada	Div. III-Va1	38° 40' 46"	9° 10' 41"	5
Palmela	Palmela	Pinhal and Castelo de Palmela Sandstones	38° 33' 56"	8° 53' 58"	220
Parque Eduardo VII	Lisboa	Div. I	38° 43' 55"	9° 09' 11"	90
Pedra da Anixa	Setúbal	Portinho da Arrábida deposits	38° 28' 41"	8 58' 21"°	5
Pedras Ninhas	Idanha a Nova	Murracha Group	39° 59' 21"	6° 57' 33"	522
Penacova	Penacova	Paleozoic (quartzite inselberge)	40° 16' 16"	8° 16' 53"	135
Penedo beach	Sesimbra	Lower-Upper Miocene	38° 27' 53"	9° 11' 32"	5
Penedo North	Sesimbra	Lower-Upper Miocene	38° 28' 03"	9° 11' 28"	5
Penedo South	Sesimbra	Lower Miocene	38° 27' 49"	9° 11' 32"	5
Penedos de Góis	Góis	Paleozoic (quartzite inselberge)	40° 07' 02"	8° 08' 43"	670
Penha Garcia	Idanha a Nova	Paleozoic (quartzite inselberge)	40° 02' 14"	7° 01' 24"	450
Pica-Galo (Trafaria)	Almada	Burdigalian	38° 39' 58"	9° 14' 23"	50
Pinhal Novo	Palmela	Plio-Pleistocene	38° 37' 53"	8° 55' 01"	50
Podence	Macedo de Cavaleiros	Aveleda Fm.	41° 35' 27"	6° 54' 38"	625
Pombal	Pombal	Serra de Sacões Group	39° 54' 50"	8° 37' 39"	75

(continued)

(continued)

Locality	Municipality	Geological unit	Latitude	Longitude W (-)	Altitude (m)
Pônsul	Idanha a Nova	Paleozoic	39° 57′ 19″	7° 11′ 52″	310
Ponte de Sôr	Ponte de Sôr	Silveirinha dos Figos Fm.	39° 15′ 01″	8° 00′ 31″	100
Portela das Necessidades	Setúbal	Benfica Fm.	38° 31′ 45″	8° 59′ 00″	170
Portinho da Arrábida	Setúbal	Lower to Middle Miocene	38° 28′ 49″	8° 58′ 27″	5
Portinho da Costa	Almada	Divs II, III, IVa, IVb	38° 40′ 32″	9° 13′ 10″	5
Porto	Porto	Paleozoic	41° 08′ 58″	8° 36′ 35″	83
Pote de Água	Lisboa	Div. IVb	38° 45′ 34″	9° 08′ 11″	100
Póvoa de Santarém	Santarém	Middle Miocene	39° 23′ 28″	8° 36′ 40″	50
Póvoa de Santo Adrião	Odivelas	Paleogene	38° 47′ 56″	9° 09′ 48″	30
Quarteira	Loulé	Ludo Fm.	37° 04′ 05″	8° 06′ 15″	5
Quelfes	Olhão	Cacela Fm.	37° 03′ 17″	7° 49′ 53″	25
Quinta da Barbuda	Lisboa	Divs. IVb, Va1, Va2, Va3, Vb	38° 44′ 57″	9° 07′ 46″	65
Quinta da Carrapata	Lisboa	Div. IVb	38° 46′ 19″	9° 10′ 00″	105
Quinta da Conceição	Lisboa	Div. Va2	38° 43′ 38″	9° 07′ 03″	50
Quinta da Farinheira	Lisboa	Div. Vb	38° 44′ 40″	9° 06′ 51″	55
Quinta da Marquesa	Alenquer	Benfica Fm.	39° 00′ 45′	8° 58′ 19″	20
Quinta da Noiva	Lisboa	Div. IVb	38° 45′ 43″	9° 07′ 55″	95
Quinta da Piedade	Lisboa	Div. I	38° 46′ 20″°	9° 11′ 02″	100
Quinta da Silvéria	Lisboa	Div. Vb	38° 46′ 02″	9° 08′ 47″	100
Quinta das Flamengas	Lisboa	Div. Vb	38° 45′ 00″	9° 07′ 05″	50
Quinta das Pedreiras	Lisboa	Divs. IVb e Va1	38° 45′ 58″	9° 09′ 24″	100
Quinta das Rosas	Almada	Divs. Vb, Vc	38° 40′ 01″	9° 11′ 51″	100
Quinta de Sta Margarida	Nave de Haver, Almeida	Vilariça arkoses	40° 30′ 12″	6° 52′ 08″	825
Quinta do Anjo	Palmela	Quinta do Anjo Sands and Marls	38° 33′ 46″	8° 55′ 25″	175
Quinta do Bacalhau	Lisboa	Div. IVb	38° 44′ 22″	9° 07′ 56″	70
Quinta do Carvalhal	Rio Maior	Middle Miocene	39° 16′ 49″	8° 47′ 33″	25
Quinta do Fidié	Lisboa	Div. IVb	38° 45′ 36″	9° 09′ 02″	95
Quinta do Narigão	Lisboa	Div. IVb	38°45′ 47″	9° 08′ 00″	110
Quinta do Pombeiro	Lisboa	Div. Va2	38° 45′ 40″	9° 07′ 43″	90
Quinta Grande	Lisboa	Div. Vb	38° 46′ 50″	9° 08′ 31″	120
Quintanelas	Sintra	Langhian	38° 50′ 42″	9° 17′ 42″	175
Rapoula	Castelo Branco	Paleozoic	39° 52′ 28″	7° 41′ 11″	315
Redondo	Redondo	Paleozoic	38° 38′ 59″	7° 32′ 53″	275

(continued)

(continued)

Locality	Municipality	Geological unit	Latitude	Longitude W (-)	Altitude (m)
Ribeira da Lage	Sesimbra	Upper Miocene	38° 29′ 13″	9° 11′ 02″	5
Rio da Prata	Sesimbra	Santa Marta Fm.	38° 29′ 15″	9° 10′ 59″	5
Rio de Moinhos	Abrantes	Torre Fm., Monfortinho Fm.	39° 28′ 40″	8° 14′ 29″	50
Rio Maior	Rio Maior	Rio Maior Fm.	39° 20′ 13″	8° 56′ 11″	75
Rocha beach	Portimão	Lagos-Portimão Fm.	37° 07′ 02″	8° 32′ 32″	0
Roussa	Pombal	Roussa Fm.	39° 54′ 57″	8° 40′ 32″	110
S. Facundo	Abrantes	Almeirim Fm.	39° 22′ 24″	8° 06′ 27″	190
Sacões butte	Góis	Sacões Group	40° 09′ 17″	7° 06′ 44″	600
Safara	Moura	Moura Basin deposits	38° 06′ 30″	7° 13′ 04″	200
Sagres	Lagos	Lagos- Portimão Fm.	37° 01′ 23″	8° 56′ 17″	25
Salcelas	Macedo de Cavaleiros	Bragança Fm.	41° 32′ 12″	6° 52′ 55″	575
Santa Eulália beach	Albufeira	Lagos-Portimão Fm.	37° 05′ 16″	8° 12′ 53″	5
Santa Marta (Corroios)	Almada	Santa Marta Fm.	38° 36′ 57″	9° 08′ 24″	25
Santa Quitéria butte	Góis	Santa Quitéria Fm.	40° 00′ 30″	8° 42′ 36″	340
Sarzedas mesa	Castelo Branco	Murracha Group	39° 50′ 47″	7° 41′ 19″	447
Seia	Seia	Paleozoic basement	40° 25′ 13″	7° 42′ 12″	500
Seixal	Seixal	Santa Marta Fm.	38° 38′ 33″	9° 06′ 22″	15
Sendas (areeiro)	Macedo de Cavaleiros	Bragança Fm.	41° 35′ 56″	6° 51′ 18″	693
Sendim (2)	Miranda do Douro	Bragança Fm.	41° 23′ 57″	6° 26′ 56″	714
Sendim, (1)	Miranda do Douro	Bragança Fm.	41° 23′ 44″	6° 26′ 27″	722
Senhor das Chagas	Alcácer do Sal	Paleozoic basement	38° 20′ 30″	8° 28′ 54″	90
Senhora da Vitória	Alcobaça	Carnide sands	39° 42′ 06″	9° 03′ 01″	0-3
Setúbal	Setúbal	Santa Marta Fm.	38° 31′ 28″	8° 53′ 35″	10
Sicó Mountain	Pombal	Mesozoic	39° 55′ 00″	8° 33′ 00″	425
Silva	Miranda do Douro	Bragança Fm	41° 30′ 44″	6° 26′ 45″	704
Silveirinha dos Figos	Castelo Branco	Silveirinha dos Figos Fm.	30° 50′ 25″	7° 40′ 21″	349
Sines	Sines	Igneous sub-volcanic	37° 57′ 05″	8° 52″ 33″	15
Sintra	Sintra	Mesozoic	38° 47′ 54″	9° 23′ 17″	200

(continued)

(continued)

Locality	Municipality	Geological unit	Latitude	Longitude W (-)	Altitude (m)
Sobreda	Carregal do Sal	Coja Fm.	40° 27′ 19″	7° 48′ 44″	380
Sortes	Sortes, Bragança	Aveleda Fm.	41° 42′ 14″	6° 48′ 43″	744
Souto da Velha	Torre de Moncorvo	Aveleda Fm.	41° 11′ 40″	6° 55′ 36″	649
Sra do Nazo	Miranda do Douro	Aveleda Fm.	41° 36′ 31″	6° 20′ 43″	812
St. Combinha	Macedo Cavaleiros	Paleozoic	41° 34′ 41″	6° 53′ 23″	650
St. Margarida	Almeida	Vilariça arkoses	40° 30′ 12″	6° 52′ 08″	825
Telhada	Góis	Telhada Fm.	40° 12′ 10″	8° 09′ 15″	324
Tomar	Tomar	Tomar Fm.	39° 36′ 13″	8° 24′ 47″	75
Tondela	Tondela	Lobão Fm.	40° 30′ 59″	8° 04′ 51″	300
Tore	Atlantic Ocean	Plio-Pleistocene	39° 25′ 09″	12° 52′ 22″	- 3500
Torre	Castelo Branco	Torre Fm.	40° 02 42″	7° 31′ 18″	500
Torre de Moncorvo	Torre de Moncorvo	Paleozoic	41° 10′ 32″	7° 03′ 09″	380
Trafaria	Almada	Burdigalian	38° 40′ 11″	9° 14′ 20″	20
Ulme	Chamusca	Ulme Fm.	39° 18′ 55″	9° 26′ 05″	50
Universidade Católica	Lisboa	Div. II	38° 44′ 56″	9° 09′ 54″	85
Vale Álvaro (distal)	IP4, Bragança	Vale Álvaro Fm.	41° 49′ 12″	6° 45′ 15″	649
Vale Álvaro (proximal)	Old Bragança railway	Vale Álvaro Fm.	41° 48′ 44″	6° 46′ 26″	700
Vale da Porca (1)	Macedo de Cavaleiros	Bragança Fm.	41° 31′ 46″	6° 54′ 42″	624
Vale da Porca (2)	Macedo de Cavaleiros	Bragança Fm.	41° 31′ 27″	6° 53′ 41″	583
Vale de Frades	Vimioso	Aveleda Fm.	41° 38′ 31″	6° 29′ 05″	743
Vale de Guizo	Alcácer do Sal	Vale de Guizo Fm.	38° 17′ 38″	8° 28′ 17″	25
Vale Furado	Alcobaça	Bom Sucesso Fm.	39° 39′ 52″	9° 03′ 49″	57
Vau	Portimão	Lagos-Portimão Fm.	37° 07′ 12″	8° 33′ 36″	5
Verín	Spain	Paleozoic	41° 56′ 26″	7° 26′ 21″	386
Vidigueira	Vidigueira	Paleozoic	38° 12′ 35″	7° 48′ 04″	220
Vila Real	Vila Real	Paleozoic	41° 17′ 45″	7° 44′ 47″	420
Vila de Rei	Vila de Rei	Paleozoic	39° 40′ 31″	8° 08′ 48″	430
Vila Nova da Rainha	Azambuja	Middle Miocene	39° 02′ 16″	8° 55′ 15″	10
Vila Nova das Patas	Mirandela	Mirandela Fm.	41° 30′ 23″	7° 11′ 11″	254

(continued)

(continued)

Locality	Municipality	Geological unit	Latitude	Longitude W (-)	Altitude (m)
Vila Velha de Ródão	Vila Velha de Ródão	Paleozoic	39° 39' 04″	7° 40' 24″	100
Vilariça	Mogadouro	Vilariça arkoses	41° 22' 41″	6° 36' 40″	800
Vimioso	Vimioso	Bragança Fm.	41° 37' 14″	6° 30' 52″	689
Xabregas	Lisboa	Div. VIa	38° 43' 41″	9° 06' 43″	15
Zavial	Vila do Bispo	Lagos-Portimão Fm.	37° 02' 47″	8° 52' 14″	10

Index

J. Pais et al., *The Paleogene and Neogene of Western Iberia (Portugal)*,   151
SpringerBriefs in Earth Sciences, DOI: 10.1007/978-3-642-22401-0,
© João Pais 2012